Atlas of Descriptive Embryology

Fourth Edition

ABOUT THE COVER:
Photomicrograph of sea urchin embryos; whole mount of fixed
and stained specimens; 16-mm achromatic objective; modified
Rheinberg illumination.

Atlas of Descriptive Embryology

Fourth Edition

Willis W. Mathews

Department of Biological Sciences
Wayne State University

Macmillan Publishing Company
New York

Collier Macmillan Publishers
London

Macmillan Publishing Company
866 Third Avenue, New York, New York 10022

Collier Macmillan Canada, Inc.

Library of Congress Cataloging in Publication Data

Mathews, Willis W.
  Atlas of descriptive embryology.

  1. Embryology—Atlases. I. Title. [DNLM: 1. Embryology
—atlases. QS 617 M429a]
  QL956.M38   1986   591.3'32'0222   85-4840
  ISBN 0-02-377140-2

Printing: 9 10 11 12 13 14 Year: 3 4 5 6

ISBN 0-02-377140-2

Preface to the Fourth Edition

This new enlarged edition contains some major additions. They include a new chapter on the development of Amphioxus, consisting of 15 photomicrographs, a selection of 25 of the classical Conklin drawings and 23 additional drawings of oocyte maturation and fertilization, gastrulation and early organ formation. This addition was advised in a questionnaire circulated among some users of the third edition. Many other helpful suggestions were received and I wish to thank all who responded for their thoughtful and informative comments. Our need to keep the book within reasonable bounds prevented the inclusion of some other recommended subjects.

Ten additional photomicrographs replace or expand the coverage of other chapters. Improved coverage was also accomplished by the addition of 29 drawings and diagrams, including 5 of gametogenesis, 4 of the sea urchin, 2 of the frog, 12 of the chick, 6 of the pig and one of the human placenta. Most of these drawings are well known because of their usefulness. Although detailed acknowledgements are given in the captions, I wish to express here my appreciation for the kind permission granted by several publishers to borrow from the works of some of their well-known authors, including B.I. Balinsky, E.G. Conklin, A.F. Huettner, O.E. Nelson, B.M. Patten, E.B. Wilson, Emil Witschi and others. The publishers of those writers include The Macmillan Publishing Co., The Blakiston Co., The McGraw-Hill Book Co., W.B. Saunders, a division of CBS Publishing, and the Wistar Institute Press. I also happily express gratitude and appreciation to my editor, Mr. Gregory Payne, Publisher, The Macmillan Publishing Co. for his encouragement and assistance.

The 100 new illustrations have required an expansion of the Glossary and Synopsis of Development section and this has been done. I hope the new material is ''on target'' and will be found helpful by many users both old and new.

W.W.M

Preface to the First Edition

Descriptive embryology still constitutes a body of knowledge fundamental to modern developmental biology. Yet, the growth of experimental and biochemical embryology and other curricular demands allow less time for the course study of descriptive embryology. As a consequence, students have generally felt the need for detailed, accurate pictures of their laboratory materials which are fully labeled. This atlas was prepared, hopefully, to satisfy such a need. With its help, together with that of his text and laboratory manual, a student should be able to carry forward his studies quickly, efficiently and mainly by his own efforts. With descriptive embryology in hand, the student and his lecturer will then be free to devote more time to comparative, physiological and experimental studies of development.

The slide preparations which were photographed for this atlas were virtually all obtained from biological supply houses. The figures should, therefore, closely resemble the slides the student will receive for his laboratory work. They will provide a useful supplement to the reconstructions, diagrams and incompletely labeled photomicrographs found in most texts and manuals.

The problem of what to include in the atlas was easily resolved in its main features. The chick embryo has been standard laboratory material for the study of development since classical times. Amphibian and sea urchin embryos have been used for many important experimental investigations of development. Some acquaintance with the development of these groups prepares the student to understand the intricacies of the experiments. The pig embryo is certainly the most widely used example of mammalian development, so it was included without hesitation. Materials for the study of gametogenesis and fertilization are available in greater variety, but the cat, the rat and *Ascaris* serve very well.

The Glossary will likely prove to be valuable. It contains all terms used in the figures. Following each term is a list of figures in which it occurs. Then, related structures are given and the term defined. Next, the development of most structures including their origin and developmental fate is summarized. Important synonyms are also listed. The definitions of terms and summaries of development are occasionally incomplete where they were necessarily limited to the animal groups included in the atlas.

If a word of advice may be offered to the student, you will do well to study your slides and other preparations thoroughly. The illustrations in this atlas are not an adequate substitute for firsthand observations of embryos. No book of practical size could show all microscopic structures and their interrelations that are important. Use the atlas to check your identifications of structures, then fully explore the morphology of the parts with your microscope. The atlas will again be helpful in reviewing for exams. The figures will provide a quick recapitulation of your slides and the Glossary will supply a summary of the development of each part.

Many hollow organs and structures have been identified on the plates of this atlas by extending a label line to the cavity or lumen of the part. This practice helps to clarify the labels, but it should be recognized by the student that the organ so identified is actually the wall or tissue surrounding the space. Examples of this kind of labeling are: blood vessel, fig. 2, archenteron, figs. 36, 87.

My wife, Vivian, prepared the drawings inserted in many of the figures. I gratefully acknowledge this and much other help in preparing this atlas.

W.W.M.

Contents

region • 96 Neural tube stage, transverse section through optic vesicles • 97 Neural tube stage, transverse section through optic placode • 98 Neural tube stage, transverse section through pharynx • 99 Neural tube stage, transverse section through nephrotome • 100 Neural tube stage, transverse section through hindgut • 101 Tail bud stage, sagittal section • 102 Tail bud stage, frontal section through pharynx.

in the pig embryo • 259 Transverse section through ductus venosus • 260 Development of the hepatic portal and umbilical veins in the pig embryo • 261 Transverse section through brachial plexus • 262 Transverse section through common bile duct • 263 Transverse section through gall bladder • 264 Reconstruction of the stomach and duodenum of the 9.4-mm pig embryo • 265 Transverse section through pancreas • 266 Transverse section through genital ridge • 267 Transverse section through urogenital sinus • 268 Transverse section through metanephros • 269 Transverse section through common iliac artery • 270 Transverse section through lumbo-sacral plexus • 271 Transverse section through spinal nerves.

1. Gametogenesis

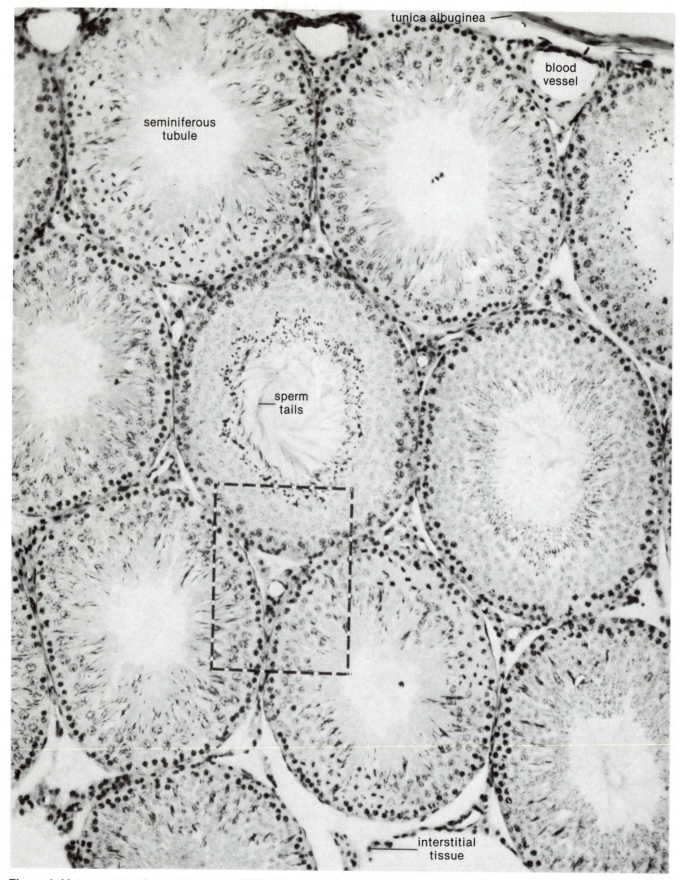

tunica albuginea

blood vessel

seminiferous tubule

sperm tails

interstitial tissue

Figure 1 Mature rat testis, section (mag. 200X). The dashed rectangle indicates the area shown in fig. 2.

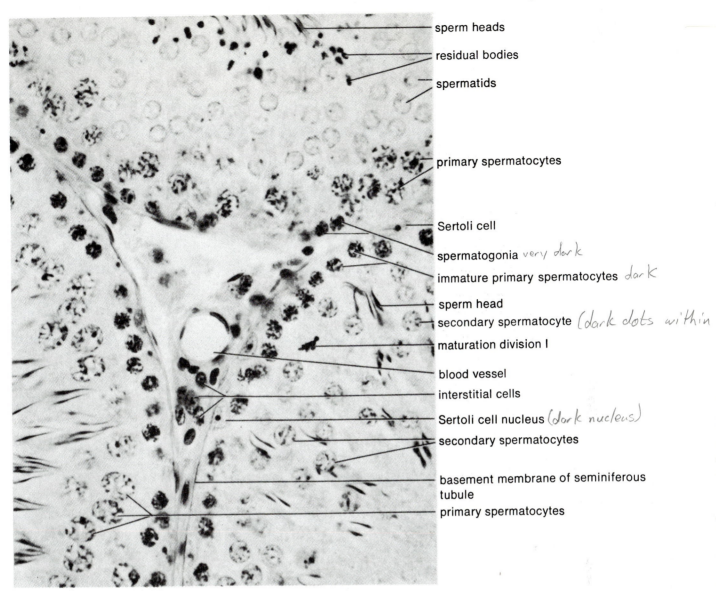

sperm heads

residual bodies

spermatids

primary spermatocytes

Sertoli cell

spermatogonia *very dark*

immature primary spermatocytes *dark*

sperm head

secondary spermatocyte *(dark dots within*

maturation division I

blood vessel

interstitial cells

Sertoli cell nucleus *(dark nucleus)*

secondary spermatocytes

basement membrane of seminiferous tubule

primary spermatocytes

Figure 2 Mature rat testis, section (mag. 650X). Label lines indicate cell nuclei. Outlines of surrounding cytoplasm are indistinct.

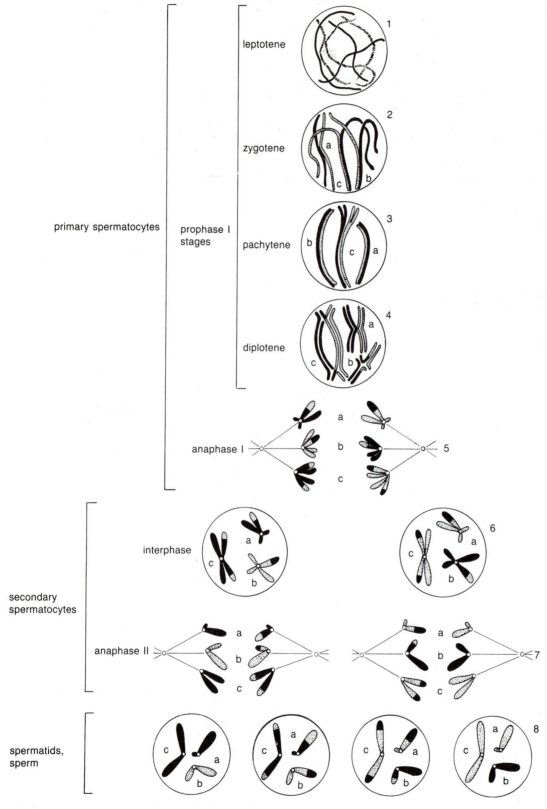

Figure 3 General diagram of meiosis, illustrating the union, separation, and distribution of the chromosomes. (From *Cell Biology*, Sixth Edition, by E. D. P. DeRobertis, Francisco A. Saez, and E. M. F. DeRobertis. Copyright © 1975 by W. B. Saunders Company. Used with permission of W. B. Saunders, a division of CBS College Publishing.)

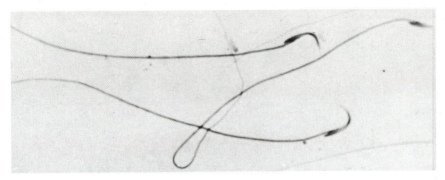

Figure 4 Rat sperm smear (mag. 710X)

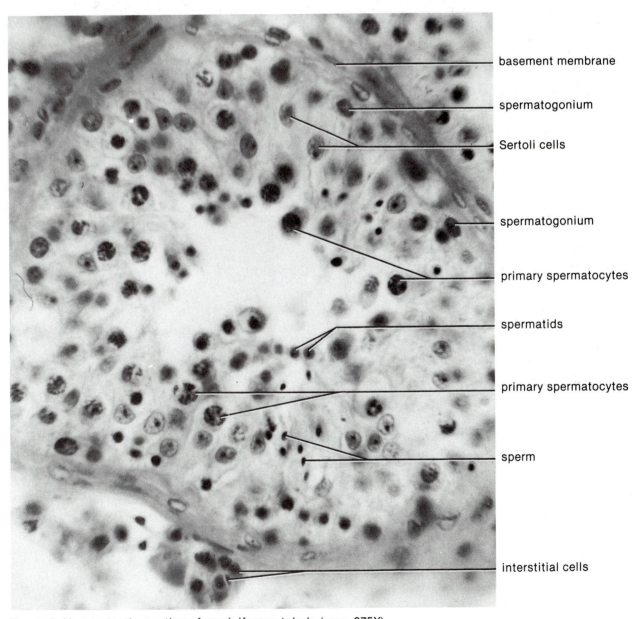

basement membrane

spermatogonium

Sertoli cells

spermatogonium

primary spermatocytes

spermatids

primary spermatocytes

sperm

interstitial cells

Figure 5 Human testis, section of seminiferous tubule (mag. 675X)

5

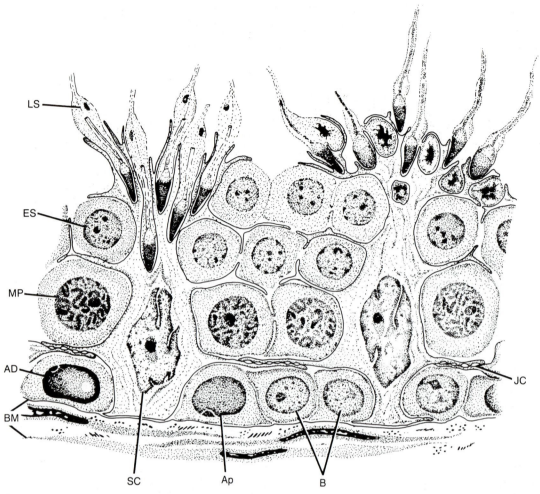

Figure 6 Diagram of a portion of a seminiferous tubule showing group of germ cells and their relationship to the Sertoli cells that extend the entire width of the tubule wall. MP, middle pachytene spermatocytes; ES, early spermatids; LS, late spermatids; AD, type A dark spermatogonium; Ap, type A pale spermatogonium; B, type B spermatogonium; BM, basement membrane; SC, Sertoli cell; JC, junctional complex. (From Karp and Berrill, *Development*, copyright 1981 by McGraw-Hill Book Company. Used with permission of McGraw-Hill Book Company. Courtesy of Y. Clermont.)

Figure 7 Mature cat ovary, section through cortex (mag. 125X)

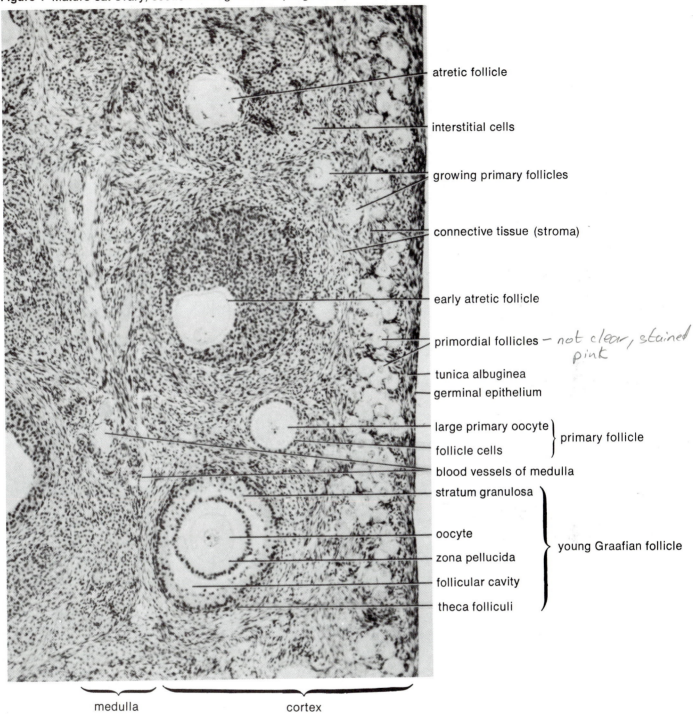

atretic follicle

interstitial cells

growing primary follicles

connective tissue (stroma)

early atretic follicle

primordial follicles — *not clear, stained pink*

tunica albuginea

germinal epithelium

large primary oocyte ⎱
⎰ primary follicle
follicle cells

blood vessels of medulla

stratum granulosa ⎱
⎰
oocyte
⎰ young Graafian follicle
zona pellucida
⎰
follicular cavity
⎰
theca folliculi ⎰

medulla   cortex

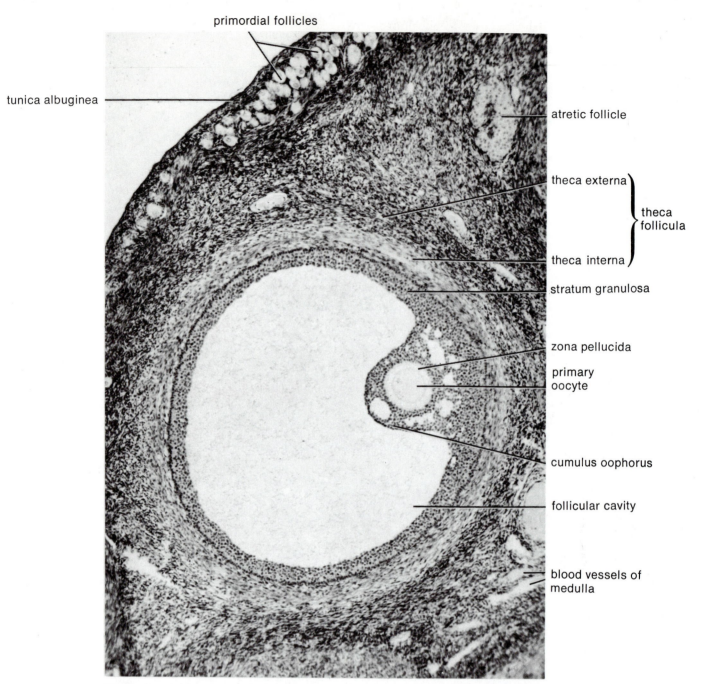

primordial follicles

tunica albuginea

atretic follicle

theca externa

theca follicula

theca interna

stratum granulosa

zona pellucida

primary oocyte

cumulus oophorus

follicular cavity

blood vessels of medulla

Figure 8 Mature cat ovary, section through large Graafian follicle (mag. 90X)

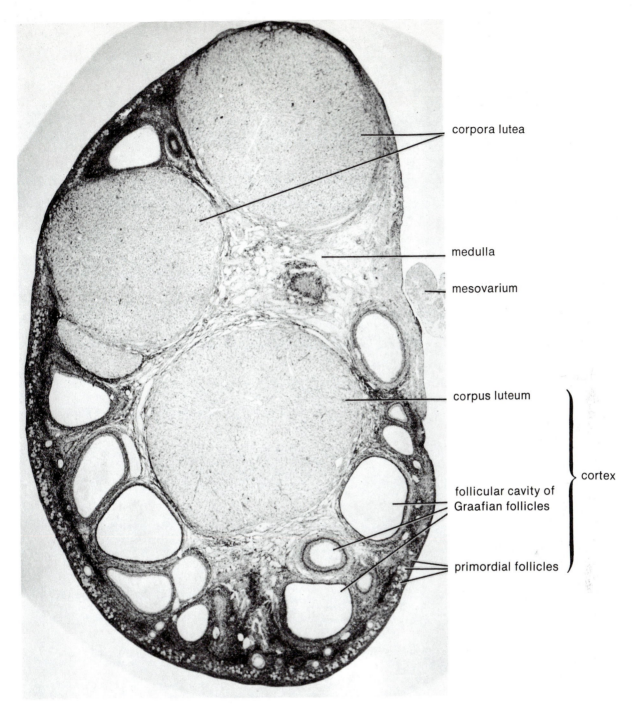

corpora lutea

medulla

mesovarium

corpus luteum

follicular cavity of Graafian follicles

primordial follicles

cortex

Figure 9 Ovary of pregnant cat, section through corpora lutea (mag. 25X)

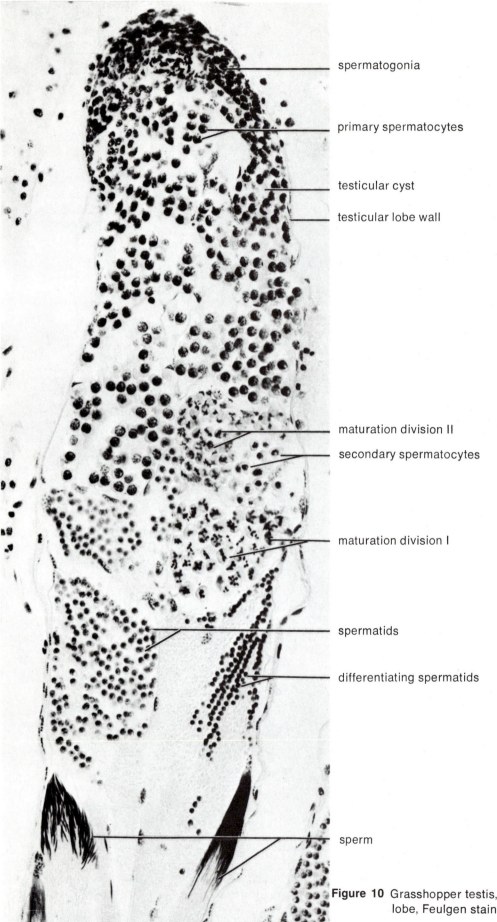

spermatogonia

primary spermatocytes

testicular cyst

testicular lobe wall

maturation division II

secondary spermatocytes

maturation division I

spermatids

differentiating spermatids

sperm

Figure 10 Grasshopper testis, longitudinal section of testicular lobe, Feulgen stain for DNA (mag. 190 X)

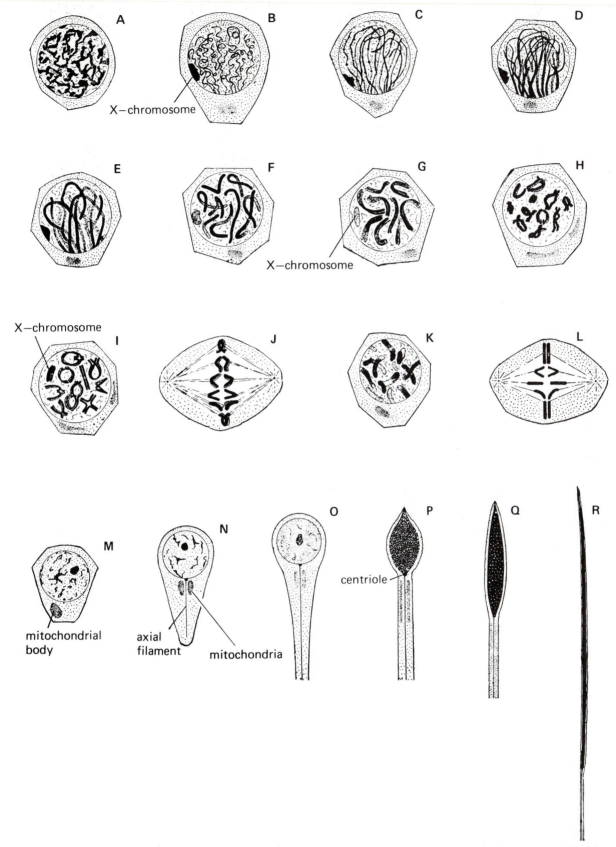

Figure 11 Spermatogenesis in the locust, *Rhomaleum tricopterum*. *a*, spermatogonium. *b–j*, primary spermatocytes. *b*, unraveling chromosomes. *c*, leptotene. *d*, zygotene. *e, f, g*, pachytene. *h*, diplotene. *i*, diakinesis. *j*, metaphase. *k, l*, secondary spermatocyte. *k*, prophase second meiosis. *l*, metaphase second meiosis. *m*, spermatid. *n–r*, spermeogeneis. (From A. F. Huettner, *Fundamentals of Comparative Embryology of the Vertebrates*. Copyright 1941 by Macmillan Publishing Company. Used with permission of Macmillan Publishing Company.)

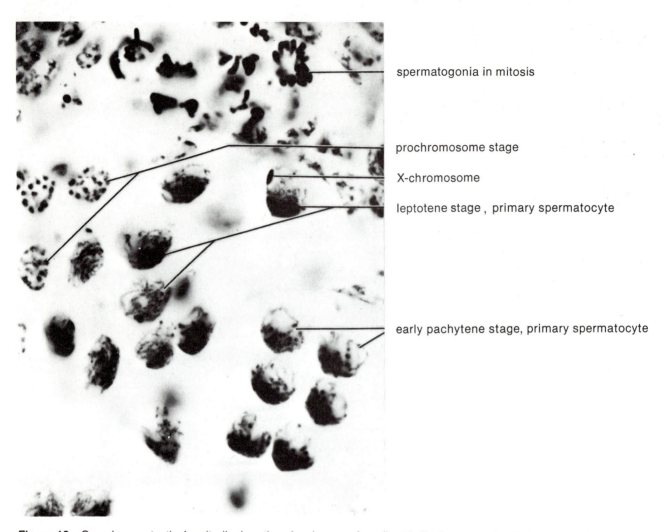

spermatogonia in mitosis

prochromosome stage

X-chromosome

leptotene stage , primary spermatocyte

early pachytene stage, primary spermatocyte

Figure 12　Grasshopper testis, longitudinal section showing area from fig. 10, Feulgen stain for DNA (mag. 880 ×)

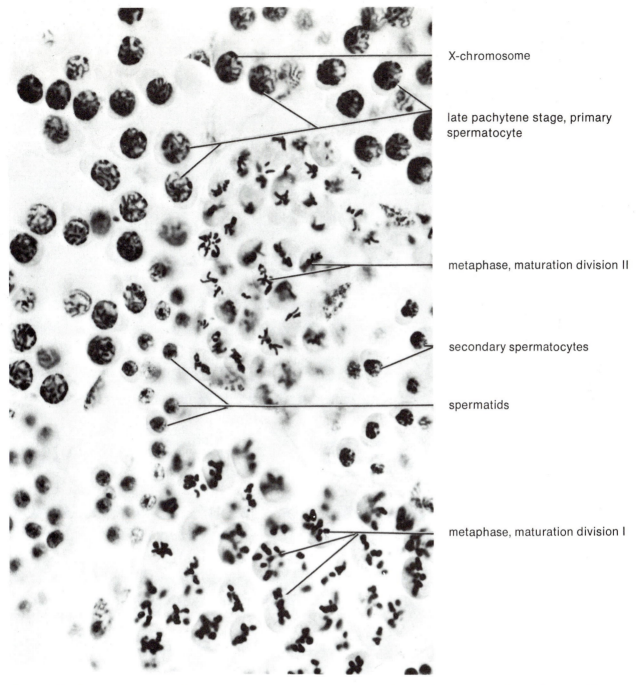

X-chromosome

late pachytene stage, primary spermatocyte

metaphase, maturation division II

secondary spermatocytes

spermatids

metaphase, maturation division I

Figure 13 Grasshopper testis, longitudinal section showing area from fig.10, Feulgen stain for DNA (mag. 550X).

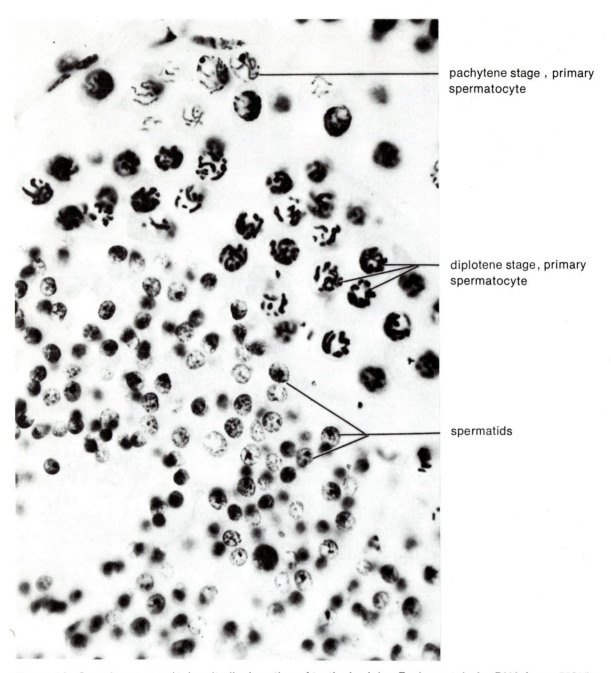

pachytene stage , primary
spermatocyte

diplotene stage, primary
spermatocyte

spermatids

Figure 14 Grasshopper testis, longitudinal section of testicular lobe, Feulgen stain for DNA (mag. 550 X)

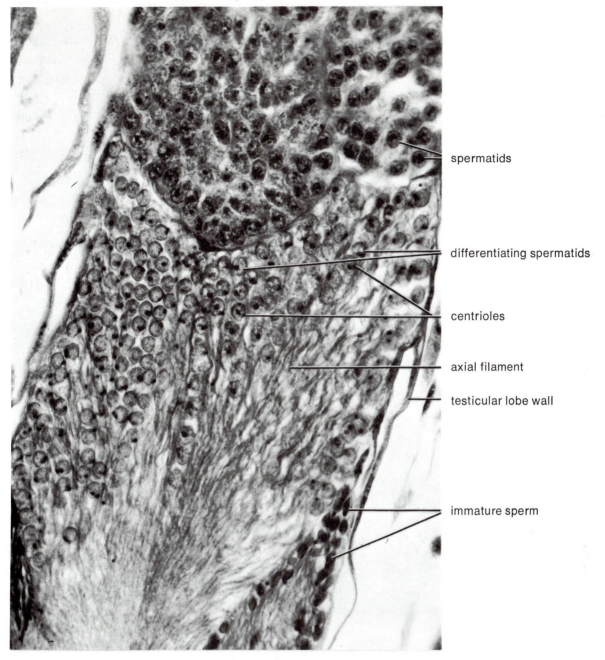

spermatids

differentiating spermatids

centrioles

axial filament

testicular lobe wall

immature sperm

Figure 15 Grasshopper testis, longitudinal section of testicular lobe, iron hematoxylin stain (mag. 550 X)

2. Maturation and Fertilization in *Ascaris*

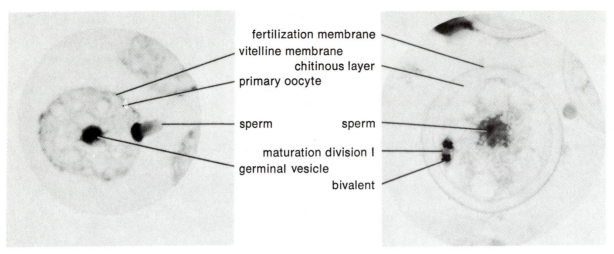

fertilization membrane
vitelline membrane
chitinous layer
primary oocyte

sperm
sperm

maturation division I
germinal vesicle
bivalent

Figure 16 Sperm penetration stage (mag. 500X)

Figure 17 First maturation division stage (mag. 500X)

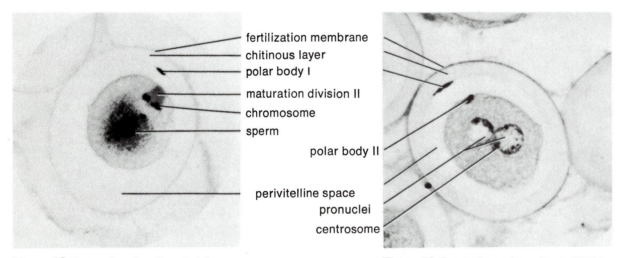

fertilization membrane
chitinous layer
polar body I
maturation division II
chromosome
sperm

polar body II

perivitelline space
pronuclei
centrosome

Figure 18 Second maturation division stage (mag. 500X)

Figure 19 Pronuclear stage (mag. 500X)

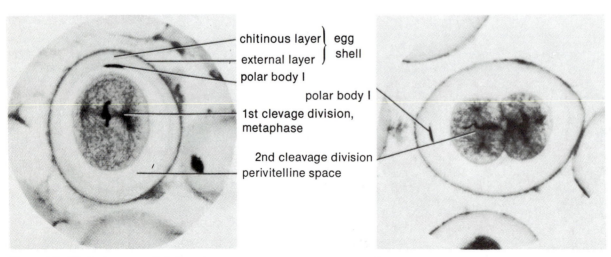

chitinous layer ⎫ egg
external layer ⎬ shell
polar body I

polar body I
1st clevage division, metaphase

2nd cleavage division
perivitelline space

Figure 20 First cleavage division stage (mag. 500X)

Figure 21 Second cleavage division stage (2-cell stage) (mag. 500X)

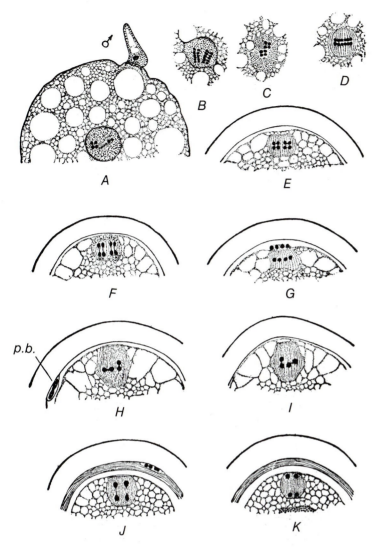

Figure 22 Formation of the polocytes (polar bodies) in *Ascaris megalocephala*, var. *bivalens* (Boveri). *a*, the egg with the sperm just entering at ♂, the germinal vesicle contains two rod-shaped tetrads (only one clearly shown), the number of chromosomes in earlier divisions having been four. *b*, the tetrads seen in profile. *c*, the same in end view. *d*, first spindle forming (in this case inside the germinal vesicle). *e*, first polar spindle. *f*, tetrads dividing. *g*, first polocyte formed containing, like the egg, two dyads. *h, i*, the dyads rotating into position for the second division, *j*, the dyads dividing. *k*, each dyad has divided into two single chromosomes, completing the reduction. *pb*, polar body I. (From E. B. Wilson *The Cell in Development and Heredity*, Copyright 1925, Macmillan Publishing Company. Used with the permission of the Macmillan Publishing Company.)

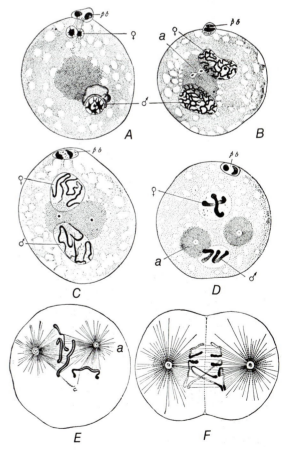

Figure 23 Fertilization of the egg of *Ascaris megalocephala*, var. *bivalens* (Boveri). *a*, the sperm has entered the egg, its nucleus is shown at ♂; above are the closing phases in the formation of the second polocyte (polar body) (two chromosomes in each nucleus). *b*, the two pronuclei (♀, ♂) in the reticular stage, the sphere (a) contains the dividing central body. *c*, chromosomes forming in the pronuclei, the central body divided. *d*, each pronucleus resolved into two chromosomes, sphere (a) double. *e*, mitotic figure forming for the first cleavage, the chromosome already split. *f*, first cleavage in progress, showing divergence of the daughter-chromosomes toward the spindle-poles (only three chromosomes shown). (From E. B. Wilson, *The Cell in Development and Heredity*, Copyright 1925, Macmillan Publishing Company. Used with the permission of Macmillan Publishing Company.)

3. Sea Urchin Development

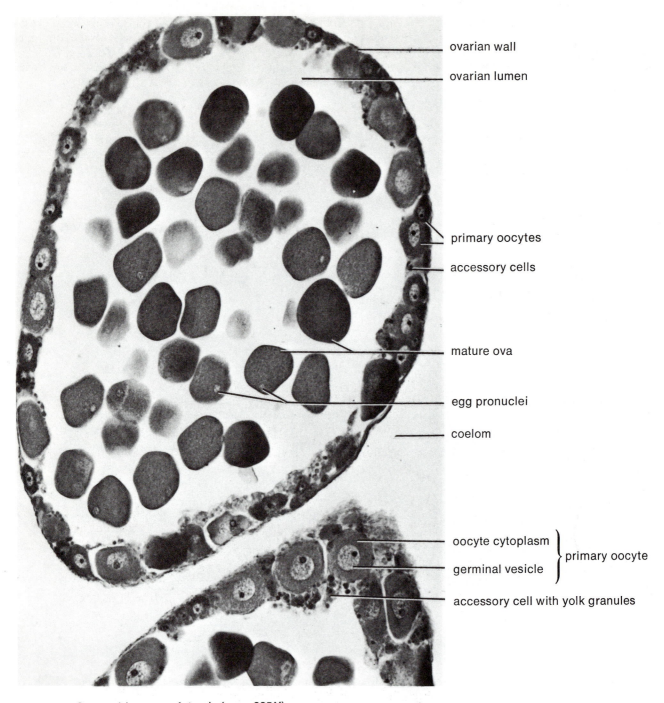

ovarian wall

ovarian lumen

primary oocytes

accessory cells

mature ova

egg pronuclei

coelom

oocyte cytoplasm

germinal vesicle

primary oocyte

accessory cell with yolk granules

Figure 24 Sea urchin ovary, *Arbacia* (mag. 225 X)

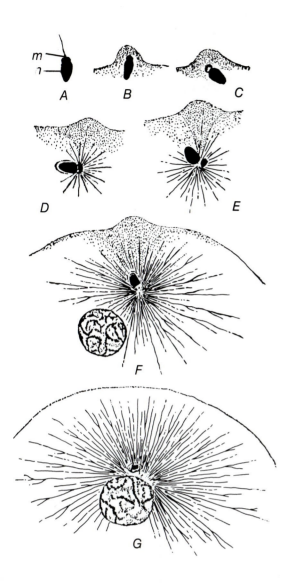

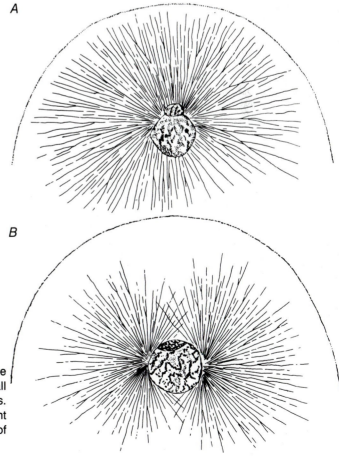

Figure 25 Entrance and rotation of the sperm-head and formation of the sperm-aster in the sea-urchin *Toxopneustes* (m, mitochondria; n, nucleus of sperm) (*a–e*, × 1600; *f, g*, × 800). (From E. B. Wilson, *The Cell in Development and Heredity*, Copyright 1925 by Macmillan Publishing Company. Used with the permission of the Macmillan Publishing Company.)

Figure 26 Conjugation of the gamete-nuclei and division of the sperm-aster in the sea-urchin *Toxopneustes,* × 1000. The small dark body is the sperm pronucleus resting upon the egg pronucleus. (From E. B. Wilson, *The Cell in Development and Heredity*, Copyright 1925 by Macmillan Publishing Company. Used with permission of Macmillan Publishing Company.)

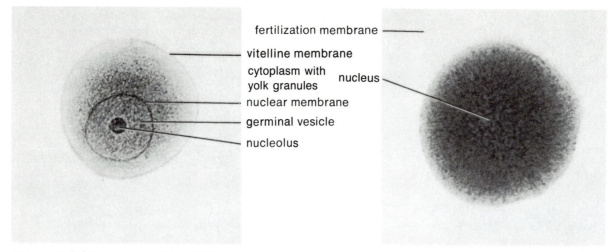

fertilization membrane ——

vitelline membrane

cytoplasm with
yolk granules   nucleus

nuclear membrane

germinal vesicle

nucleolus

Figure 27 Primary oocyte (mag. 300X)

Figure 28 Fertilized egg (mag. 300X)

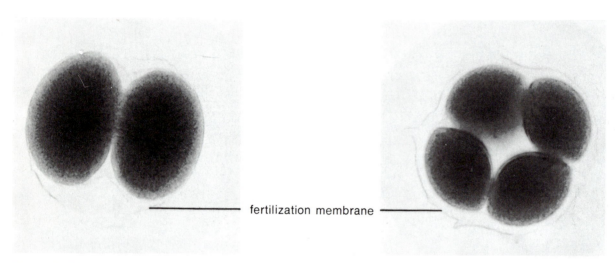

fertilization membrane ————

Figure 29 Two-cell stage (mag. 300X)

Figure 30 Four-cell stage (mag. 300X)

fertilization membrane ————

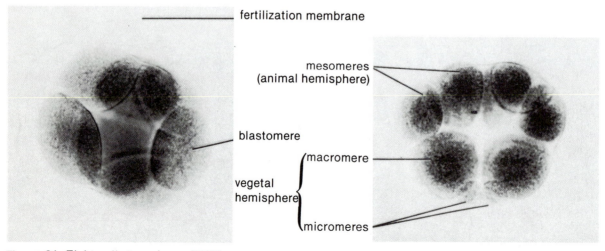

mesomeres
(animal hemisphere)

blastomere

macromere

vegetal
hemisphere

micromeres

Figure 31 Eight-cell stage (mag. 300X)

Figure 32 Sixteen-cell stage (mag. 300X)

24

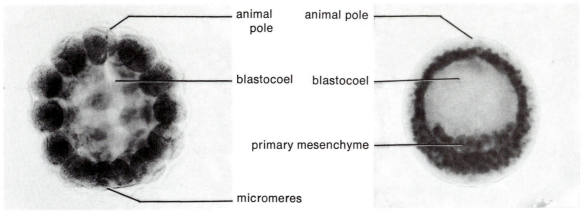

Figure 33 Early blastula stage
(mag. 300X)

Figure 34 Late blastula stage
(mag. 300X)

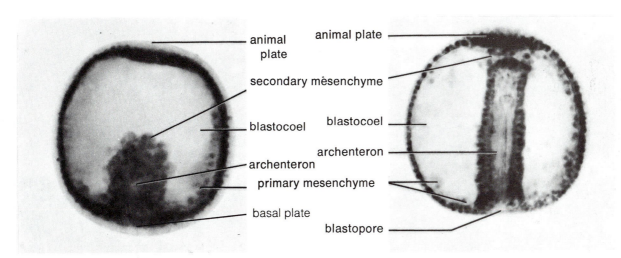

Figure 35 Early gastrula stage
(mag. 300X)

Figure 36 Late gastrula stage
(mag. 300X)

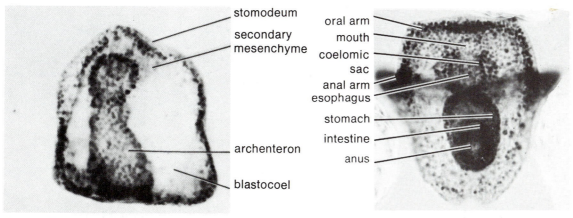

Figure 37 Prism stage
(mag. 300X)

Figure 38 Early pluteus larval stage
(mag. 300X)

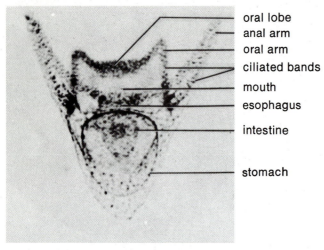

oral lobe

anal arm

oral arm

ciliated bands

mouth

esophagus

intestine

stomach

Figure 39 Late pluteus larval stage,
ventral view (mag. 200X)

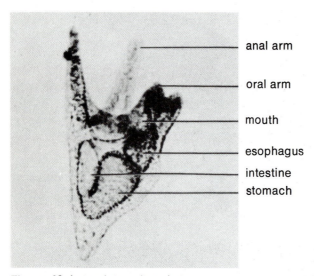

anal arm

oral arm

mouth

esophagus

intestine
stomach

Figure 40 Late pluteus larval stage,
lateral view (mag. 200X)

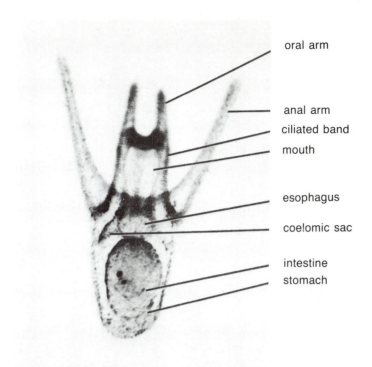

oral arm

anal arm

ciliated band

mouth

esophagus

coelomic sac

intestine

stomach

Figure 41 Mature pluteus of *Arbacia* (mag. 200 ×).

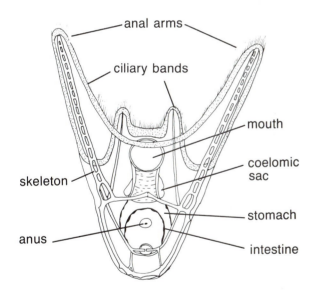

anal arms

ciliary bands

mouth

coelomic sac

skeleton

stomach

anus

intestine

Figure 42 Pluteus, larva of a sea urchin (*Tripneustes gratilla*). (From *An Introduction to Embryology*, Fifth Edition by B. I. Balinsky, assisted by B. C Babian. Copyright © 1981 by CBS College Publishing. Reprinted by permission of W. B. Saunders, a division of CBS Publishing.)

4. Starfish Development

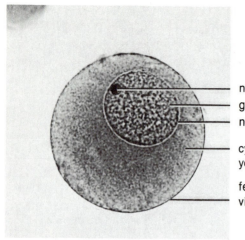

nucleolus
germinal vesicle
nuclear membrane

cytoplasm with
yolk granules

fertilization membrane
vitelline membrane

Figure 43 Primary oocyte (mag. 300X)

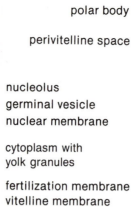

polar body
perivitelline space

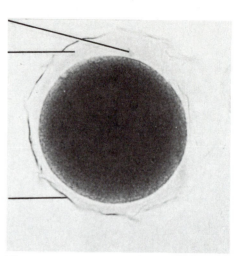

Figure 44 Fertilized egg (mag. 300X)

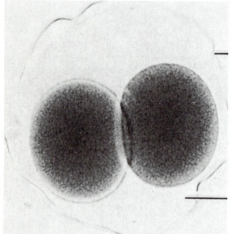

fertilization membrane

perivitelline space

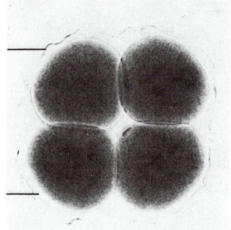

Figure 45 Two-cell stage (mag. 300X)

Figure 46 Four-cell stage (mag. 300X)

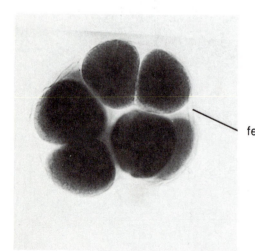

early blastocoel

fertilization membrane

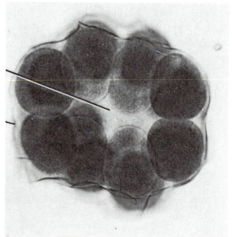

Figure 47 Eight-cell stage (mag. 300X)

Figure 48 Sixteen-cell stage (mag. 300X)

30

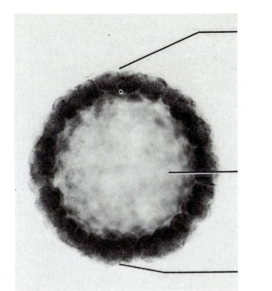

animal pole

blastocoel

vegetal pole

Figure 49 Early blastula (mag. 300X)

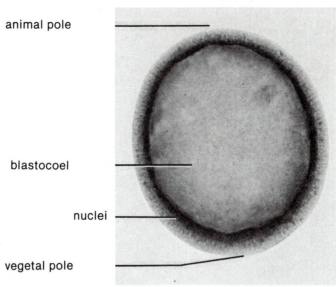

nuclei

Figure 50 Late blastula (mag. 300X)

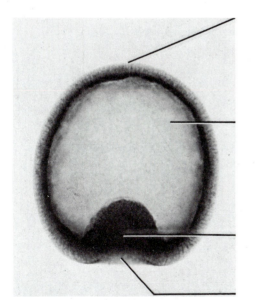

animal pole

Figure 51 Early gastrula (mag. 300X)

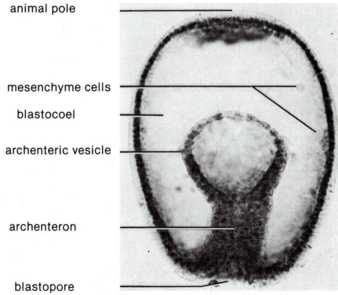

mesenchyme cells

blastocoel

archenteric vesicle

archenteron

blastopore

Figure 52 Gastrula stage (mag. 300X)

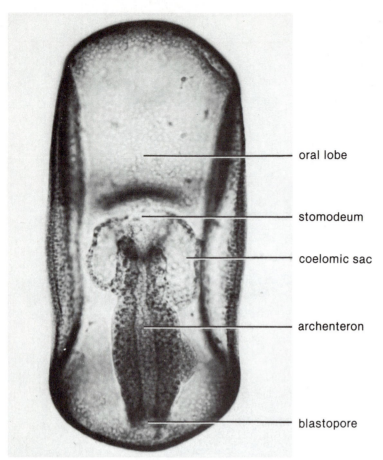

oral lobe

stomodeum

coelomic sac

archenteron

blastopore

Figure 53 Late gastrula, ventral view (mag. 300X)

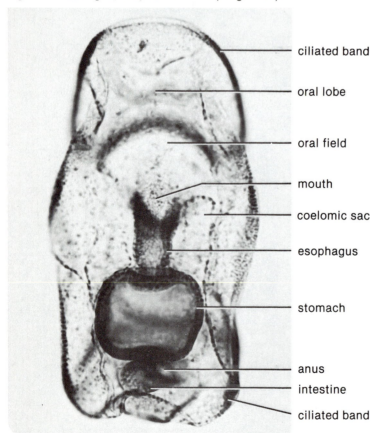

ciliated band

oral lobe

oral field

mouth

coelomic sac

esophagus

stomach

anus

intestine

ciliated band

Figure 54 Bipinnaria larva, ventral view (mag. 300X)

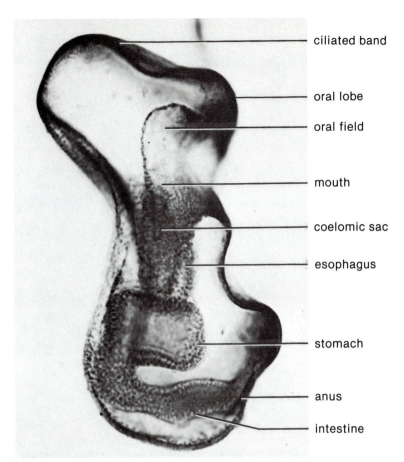

ciliated band

oral lobe

oral field

mouth

coelomic sac

esophagus

stomach

anus

intestine

Figure 55 Bipinnaria larva, lateral view (mag. 300X)

5. Development of Amphioxus

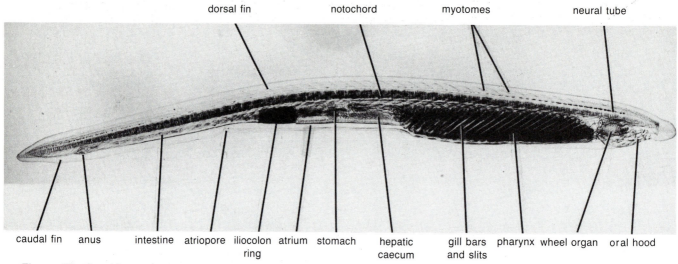

dorsal fin   notochord   myotomes   neural tube

caudal fin   anus   intestine   atriopore   iliocolon   atrium   stomach   hepatic   gill bars   pharynx   wheel organ   oral hood
ring   caecum   and slits

Figure 56   Amphioxus, immature adult, whole mount (mag. 28 ×).

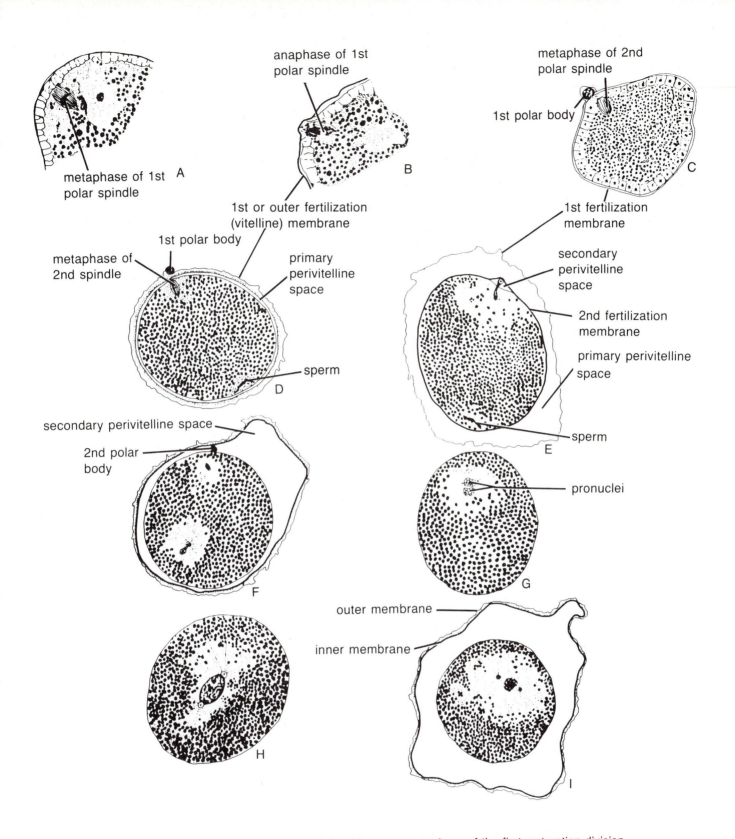

Figure 57 Fertilization and maturation of the egg of *Amphioxus. a,* metaphase of the first maturation division before sperm entrance. *b,* anaphase of the first maturation division before sperm entrance. *c,* first polar body and metaphase of the second maturation before sperm entrance. *d,* sperm penetrated egg near its vegetal pole. *e,* outer vitelline membrane separated from egg, second polar body forming. *f,* inner vitelline membrane lifted and fused with the first to form the fertilization membrane, pronuclei of egg and sperm formed, sperm-aster and second polar body present. *g,* sperm and egg pronuclei have met. *h,* pronuclei fused. *i,* prophase of first cleavage mitosis. (From O. E. Nelson, *Comparative Embryology of the Vertebrates* [after Cerfontaine and Sobatta]. Copyright 1953 by The Blakiston Co. Used with permission of McGraw-Hill Book Company.)

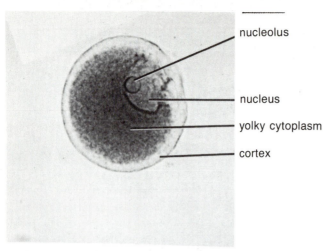

nucleolus

nucleus

yolky cytoplasm

cortex

Figure 58 Amphioxus oocyte, whole mount (mag. 250 ×).

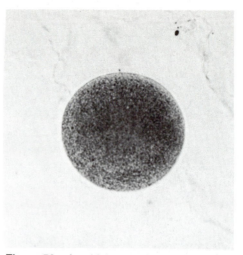

Figure 59 Amphioxus uncleaved egg, whole mount (mag. 250 ×).

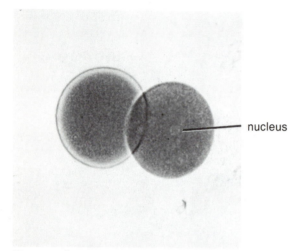

nucleus

Figure 60 Amphioxus embryo, two-cell stage, whole mount (mag. 250 ×).

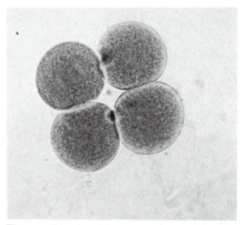

Figure 61 Amphioxus embryo, four-cell stage, whole mount (mag. 250 ×), cells are in mitosis, consequently nuclei are not visible.

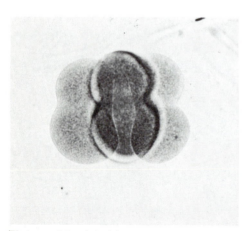

Figure 62 Amphioxus embryo, third cleavage stage, whole mount (mag. 250 ×). The third cleavage is incomplete in the telophase and is unequal.

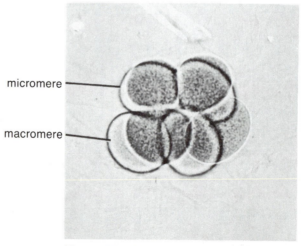

micromere

macromere

Figure 63 Amphioxus embryo, eight-cell stage whole mount (mag. 250 ×). The smaller micromeres constitute the animal hemisphere and will form ectoderm; the larger macromeres will form mesoderm and endoderm.

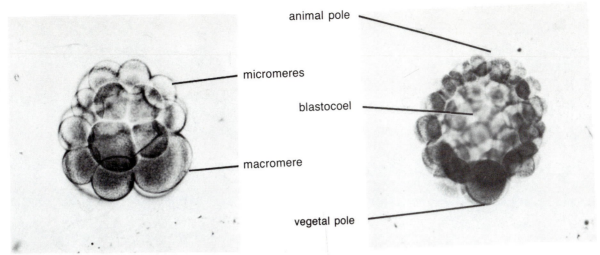

micromeres

macromere

Figure 64 Amphioxus embryo, thirty-two cell stage, whole mount (mag. 250 ×).

animal pole

blastocoel

vegetal pole

Figure 65 Amphioxus embryo, early blastula stage in optical section, whole mount (mag. 250 ×).

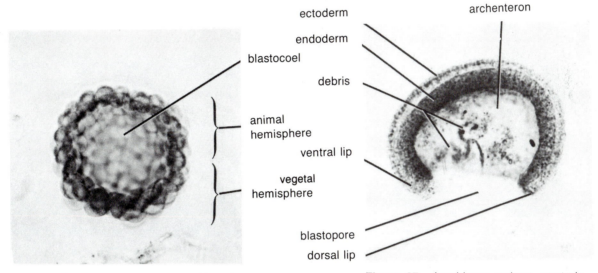

blastocoel

animal hemisphere

vegetal hemisphere

Figure 66 Amphioxus embryo, blastula stage in optical section, whole mount (mag. 250 ×). The larger vegetal cells will invaginate to form endoderm.

ectoderm

endoderm

debris

ventral lip

blastopore

dorsal lip

archenteron

Figure 67 Amphioxus embryo, gastrula stage in optical section, whole mount (mag. 250x). The invaginated endoderm and mesoderm have filled the blastocoel.

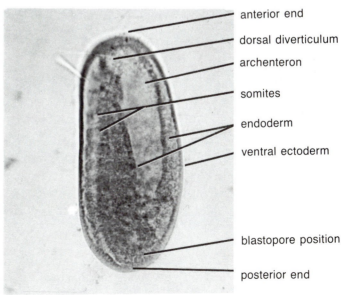

anterior end

dorsal diverticulum

archenteron

somites

endoderm

ventral ectoderm

blastopore position

posterior end

Figure 68 Amphioxus, late embryo stage in optical section, whole mount (mag. 250 ×). Gastrulation is completed and the blastopore nearly closed; embryo elongating; notochord and somites separating from the archenteron.

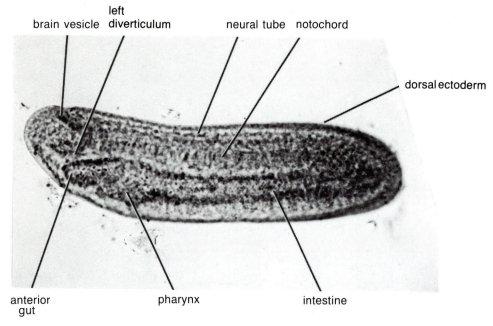

left
brain vesicle   diverticulum   neural tube   notochord

dorsal ectoderm

anterior
gut   pharynx   intestine

Figure 69 Amphioxus early larva about 26 hours after fertilization, whole mount
(mag. 250 ×).

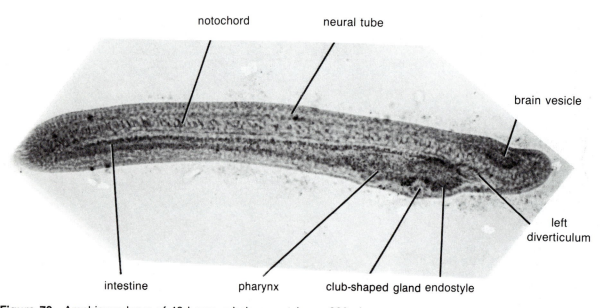

notochord   neural tube

brain vesicle

left
diverticulum

intestine   pharynx   club-shaped gland   endostyle

Figure 70 Amphioxus larva of 48 hours, whole mount (mag. 200 ×).

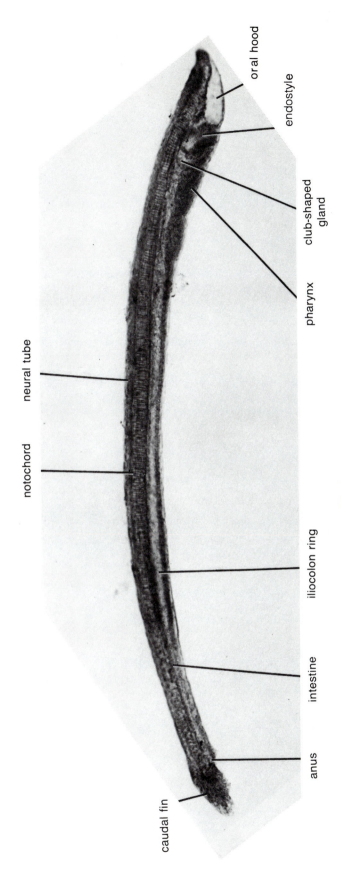

Figure 71 Amphioxus larva of 96 hours, whole mount (mag 175×).

oral hood

endostyle

club-shaped gland

pharynx

neural tube

notochord

iliocolon ring

intestine

anus

caudal fin

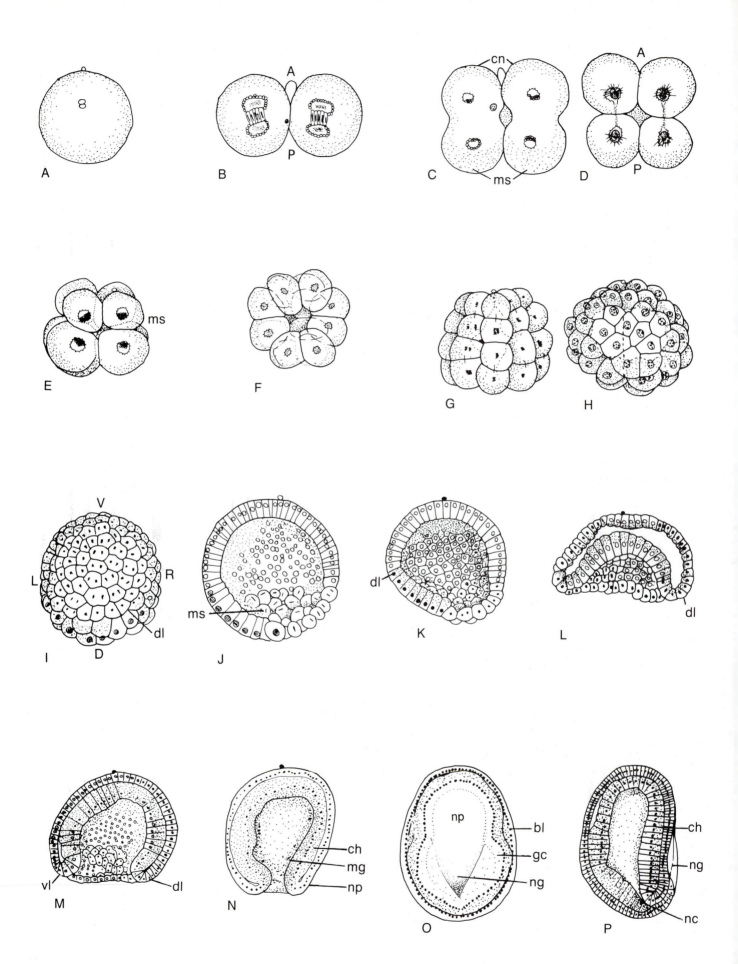

Figure 72 [opposite] Amphioxus development, fertilized egg to late embryo stages. (From E. G. Conklin; The Embryology of *Amphioxus. Jour. of Morph.* 54:69(1932). Used with the permission of the Wistar Institute Press.)

a, egg one hour after fetilization; the two pronuclei are just above center and the second polar body marks the animal pole. The dark-staining material in the vegetal region marks the mesodermal crescent. *b*, the 2-cell stage with second cleavage in progress 2 hours after fertilization. *c*, second cleavage in telophase 2 hours after fertilization; the mesodermal cresent (ms) is in the posterior region and the chorda-neural crescent opposite. *d*, vegetal view of the 4-cell stage 2 hours after fertilization; the anterior (A) cells are slightly larger. *e*, 2½ hours after fertilization the third cleavage has divided the cells unequally yielding 4 small micromeres and 4 larger macromeres; the posterior macromeres contain the mesodermal crescent (ms); the arrow indicates the body axis of the future embryo. *f*, the 16-cell stage 2½ hours after fertilization; a definite blastocoel has formed. *g*, the 32-cell stage 2½ hours after fertilization; the blastocoel is open at vegetal pole. *h*, the 64-cell stage 3¼ hours after fertilization; the vegetal cells are larger than animal cells; the blastocoel is now closed. *i*, the 256-cell stage 4 hours after fertilization; the lower large cells will form endoderm with the dorsal lip (dl) at their border. *j*, the blastula 5½ hours after fertilization seen in optical section, the small dividing cells at the posterior surface were derived from the mesodermal crescent (ms); the large posterior cells are prospective endoderm; the second polar body marks the animal pole. *k*, a blastula 5½ hours after fertilization in optical section; the endodermal cells have flattened with the future dorsal lip (dl) at one margin and the mesodermal crescent cells at the other. *l*, early gastrula about 8 hours after fertilization in optical section; the endoderm has invaginated into the blastocoel; the dorsal lip (dl) of the blastopore is on the right and contains cells of the future notochord. *m*, gastrula of 11 hours in optical section; the blastocoel has been nearly filled by the invaginated archenteron; the dorsal lip (dl) of the blastopore is on the right; mesodermal cells are still involuting over the ventral lip (vl). *n*, gastrula of 13 hours in optical section; embryo is elongating and the blastopore constricting; the dorsal wall of the embryo on the right consists of the ecodermal neural plate (np) underlain by notochord (ch) and mesodermal groove (mg), both now in the wall of the archenteron. *o*, embryo of 15-16 hours seen from dorsal side; the flattened area is the neural plate (np) depressed posteriorly as neural groove (ng); a remnant of the blastocoel (bc) persists; the margin of anchenteron (gastrocoel) (gc) is indicated by a dotted line; the blastopore is nearly closed and the neural groove is being covered by ectodermal folds derived from the ventral and lateral lips of the blastopore. *p*, embryo of 15 hours showing the ectodermal overgrowth of the neural groove (ng); a neurentric canal (nc) connects the neural groove with the blastopore; the notochord (ch) underlies the neural groove.

Key to Abbreviations, Figs. 72, 73

A, anterior
a, anus
ba, branchial rudiment
bc, blastocoel
br, branchial region
bv, brain vesicle
cg, club-shaped gland
cgd, duct of club-shaped gland
ch, notochord
cn, chorda-neural crescent
cz, ciliated zone of gut
D, dorsal side
dd, dorsal diverticulum
dl, dorsal lip
es, endostyle
gc, gastrocoel (archenteron)
lg, 1st gill slit

L., left side
ld, left diverticulum
mg, mesodermal groove
ms, mesodermal crescent
nc, neurenteric canal
ng, neural groove
np, neural plate
p, neuropore
pg, pigment granules
P, posterior
pp, preoral pit
ps, pigment spot
rd, right dorsal diverticulum
R, right side
s, somite
V, ventral side

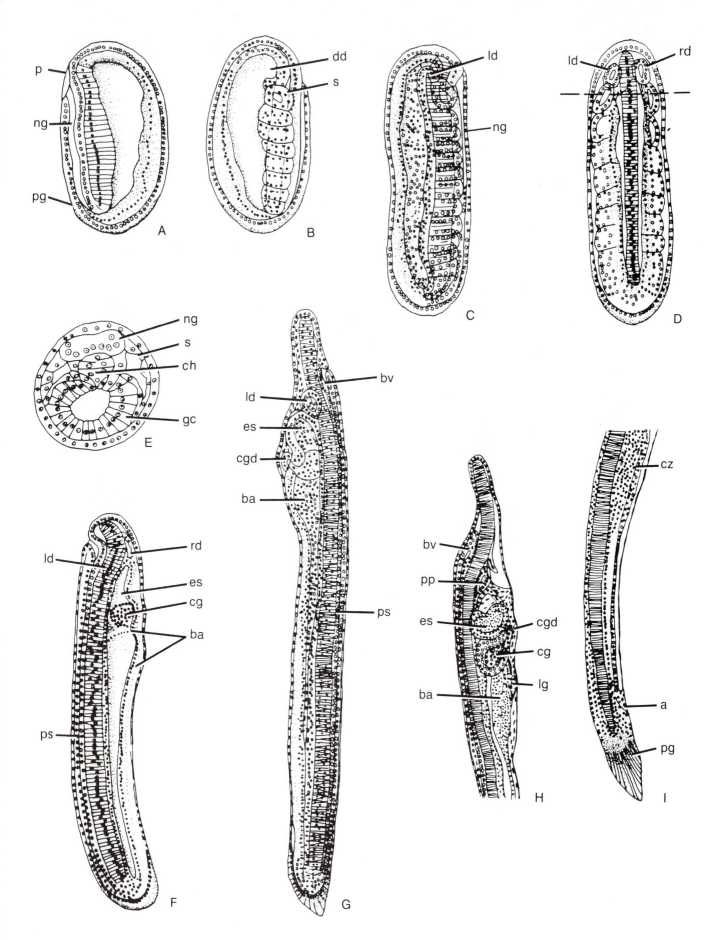

Figure 73 [opposite] Amphioxus development, late embryo to larval stages. (From E. G. Conklin; The Embryology of *Amphioxus; Jour. of Morph*. 54:69 (1932). Used with permission of the Wistar Institute Press.)

a, embryo of 16 hours viewed from right side; the ectodermal overgrowth of the neural groove (ng) is complete except for the open neuropore (p) at the anterior end of the neural groove (ng); pigment granules (pg) darken the posterior surface. *b*, embryo of 18 hours viewed from left side; a series of paired somites (s) has evaginated from the archenteron as has a dorsal diverticulum (dd) at its anterior end. *c*, embryo of 24 hours viewed from left side; elongation is marked with 10 or 11 somites; the posterior half of the neural groove (ng) has closed forming neural tube; a left dorsal diverticulum (ld) has evaginated from the archenteron; the embryo is now ciliated and hatches at this stage. *d*, embryo of 24 hours viewed from dorsal side; the central rod is the notochord bordered on both sides by a row of somites; the most anterior vesicles are the right and left diverticula (rd, ld) of the archenteron; the transverse line indicates the plane of section through the first somites shown in fig 73-E. *e*, transverse section of a 24 hour embryo though the first somites as indicated in fig. 73-D; the notochord (ch) and somites (s) have separated from the archenteron (gastrocoel) (gc); the neural groove (ng) overlies the notochord and somites. *f*, larva of 26 hours viewed from right side; the right diverticulum (rd) is expanding to form the head cavity; the left diverticulum (ld) is visible through the notochord; the rudiment of the endostyle (es) lies on the floor of the anterior gut; anterior to the branchial segment (pharynx, ba) of the gut is the club-shaped gland (cg); a light sensitive pigment spot (ps) has formed in the neural tube. *g*, larva of 48 hours viewed from the left; the notochord now extends well beyond the brain vesicle (bv); the left diverticulum (ld), endostyle (es), branchial segment (pharynx) (ba) and pigment spot (ps) are as in fig. 73-F; the club-shaped gland duct (cgd) is visible. *h*, anterior third of 96 hour larva viewed from the right; above the notochord lies the brain vesicle (bv) and below is the preoral pit (pp) derived from the left diverticulum; just posterior is the first segment of the gut into which the mouth opens from the left side and is not visible here; a dark heart-shaped endostyle (es) and the club-shaped gland follows; below the branchial segment (pharynx) (ba) is the first gill slit (lg). *i*, posterior third of 96 hour larva; a dark ciliated zone (cz) (ileocolon ring) of the intestine is forming; the intestine terminates at the anus (a) on the ventral side just anterior to the caudal fin.

Key to Abbreviations, Figs. 72, 73

A, anterior
a, anus
ba, branchial rudiment
bc, blastocoel
br, branchial region
bv, brain vesicle
cg, club-shaped gland
cgd, duct of club-shaped gland
ch, notochord
cn, chorda-neural crescent
cz, ciliated zone of gut
D, dorsal side
dd, dorsal diverticulum
dl, dorsal lip
es, endostyle
gc, gastrocoel (archenteron)
lg, 1st gill slit

L, left side
ld, left diverticulum
mg, mesodermal groove
ms, mesodermal crescent
nc, neurenteric canal
ng, neural groove
np, neural plate
p, neuropore
pg, pigment granules
P, posteror
pp, preoral pit
ps, pigment spot
rd, right dorsal diverticulum
R, right side
s, somite
V, ventral side

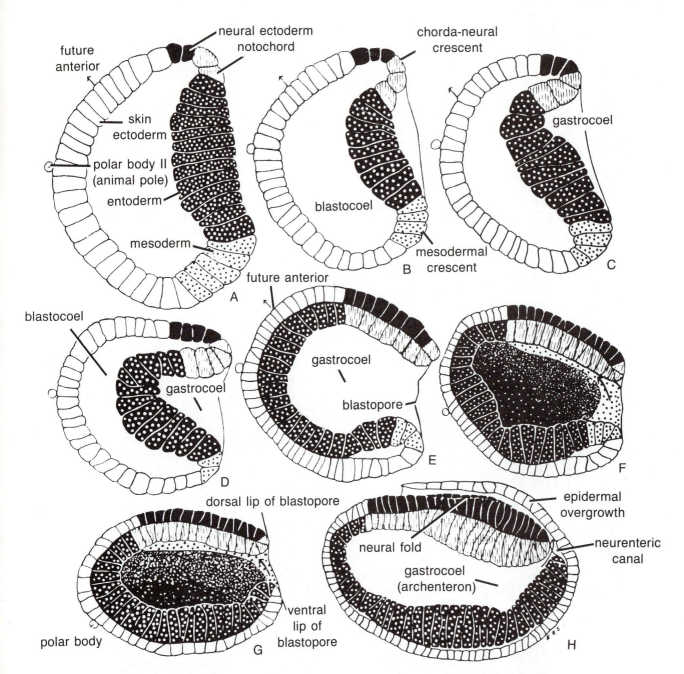

Figure 74 Gastrulation in *Amphioxus* showing positions of organ-forming areas. *a*, beginning gastrulation; animal hemisphere of epidermal and neural ectoderm, vegetal hemisphere of notochord, endoderm, and mesoderm beginning to invaginate. *b*, dorsal crescent of prospective notochord is in dorsal lip of blastopore and ventral crescent of mesoderm in ventral and lateral lips, *c* and *d*, invagination continues. *e*, gastrulation completed nearly eliminating the blastocoel and bringing ectoderm into broad contact with archenteron; neural area is underlain by notochord. *f* and *g*, as embryo elongates and blastopore constricts, the mesoderm splits and migrates anteriorly along either side of the notochord. *h*, gastrulation completed; neural ectoderm thickens to form neural plate and neural folds; ectoderm from ventral lip overgrows the neural organ forming the neurentric canal. (From O. E. Nelson, *Comparative Embryology of the Vertebrates.* After Cerfontain and Sobotta. Copyright 1953 by the Blakiston Co. Used with permission of McGraw-Hill Book Company.)

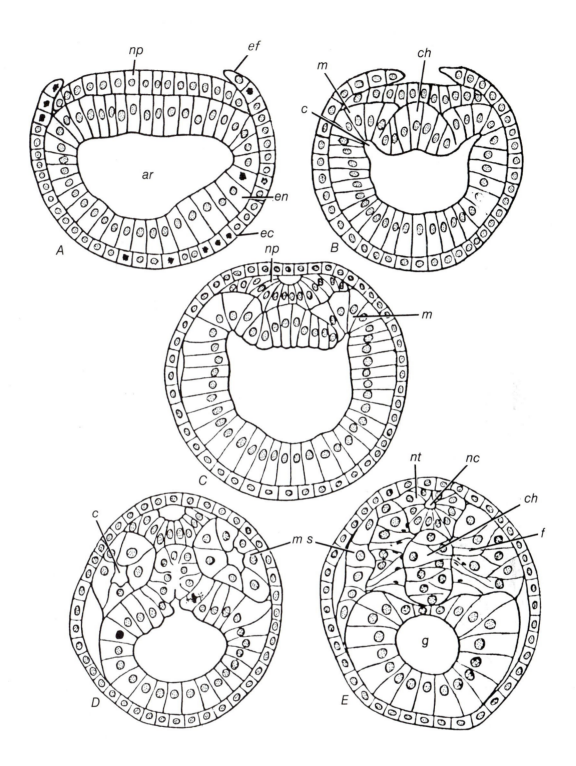

Figure 75 Transverse sections through *Amphioxus* embryos showing early organ formation. *a,* formation of neural plate and ectodermal folds from dorsal ectoderm. *b,* somites and notochord beginning to evaginate from archenteron, neural plate being covered by ectodermal folds. *c,* evagination of somites proceeds forming hollow vesicles; neural plate folds longitudinally into neural groove. *d,* first somites now separated from archenteron and possess an entrocoel cavity. *e,* section through middle somites of a 9-somite larva; notochord separated from archenteron; dorsal edges of archenteron fused and neural groove closed into neural tube; the medial wall of the somites (myotome) is forming muscle fibrillae. Abbreviations: ar, archenteron (gastrocoel); c, entercoel; ch, notochord; ec, ectoderm; ef, ectodermal fold; en, endoderm; f, muscle fibrillae; g, gut cavity; m, mesoderm; ms, mesodermal somite; nc, neural canal; mp, neural plate; nt neural tube.

(From W. E. Kellicott. *A Textbook of General Embryology.* Copyright 1913 by Henry Holt Co., after Cerfontaine.)

6. Early Development of the Frog

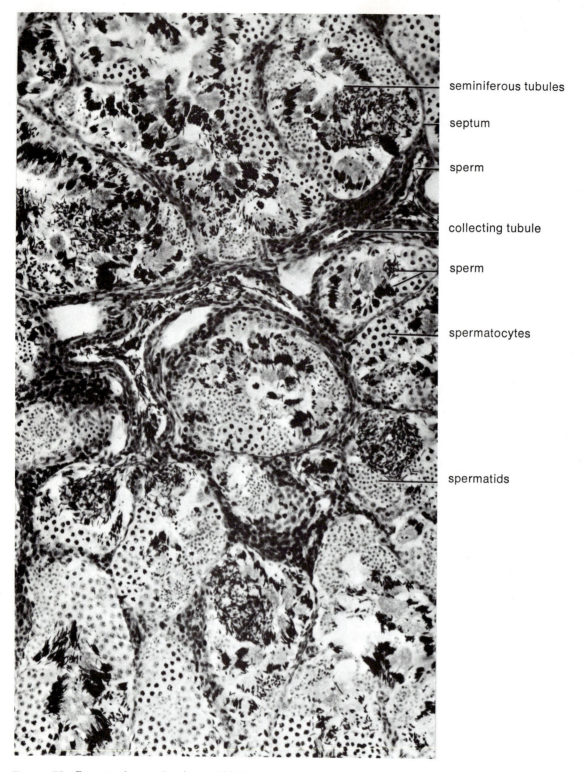

seminiferous tubules

septum

sperm

collecting tubule

sperm

spermatocytes

spermatids

Figure 76   Frog testis, section (mag. 180 X)

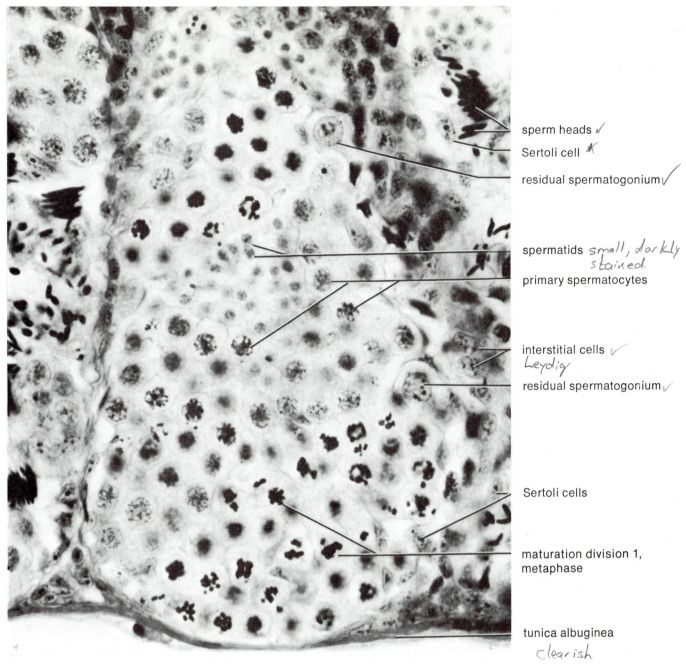

sperm heads ✓
Sertoli cell ✗
residual spermatogonium ✓

spermatids *small, darkly stained.*
primary spermatocytes

interstitial cells ✓
Leydig
residual spermatogonium ✓

Sertoli cells

maturation division 1, metaphase

tunica albuginea
clearish

Figure 77 Frog testis, section (mag. 725 X)

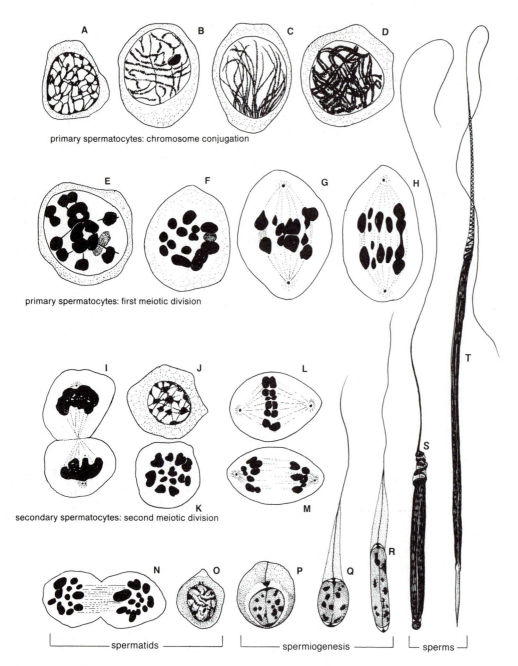

Figure 78 Spermatogenesis in the frog: Maturation phase. Drawings of single cells arranged in a progressive series, from spermatogonium just transforming into a spermatocyte (A), to mature sperms; the short and blut sylvatica type (S) is more common than the pointed temporaria type (T); × 1900. (From *Development of Vertebrates*, by Emil Witschi. Copyright © 1956 by W. B. Saunders Company. Used with permission of W. B. Saunders, a division of CBS College Publishing.)

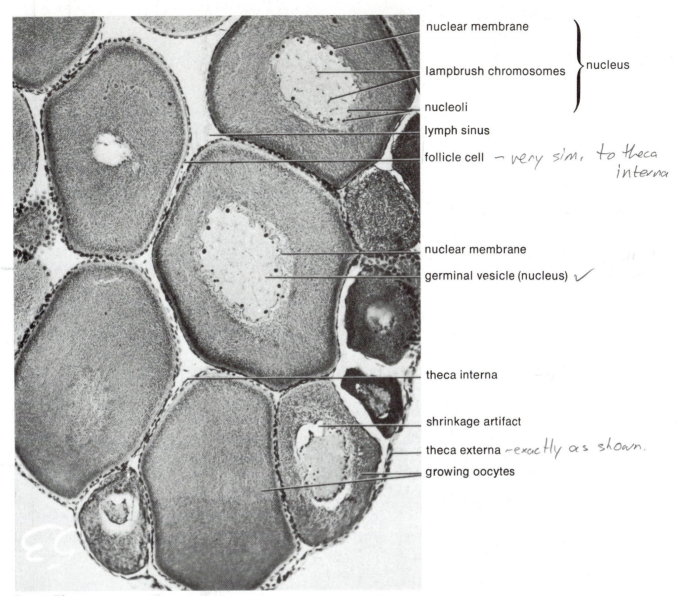

nuclear membrane ⎫
lampbrush chromosomes ⎬ nucleus
nucleoli ⎭

lymph sinus

follicle cell — very sim. to theca interna

nuclear membrane

germinal vesicle (nucleus) ✓

theca interna

shrinkage artifact

theca externa — exactly as shown.

growing oocytes

Figure 79 Frog ovary, section showing growing oocytes (mag. 135X)

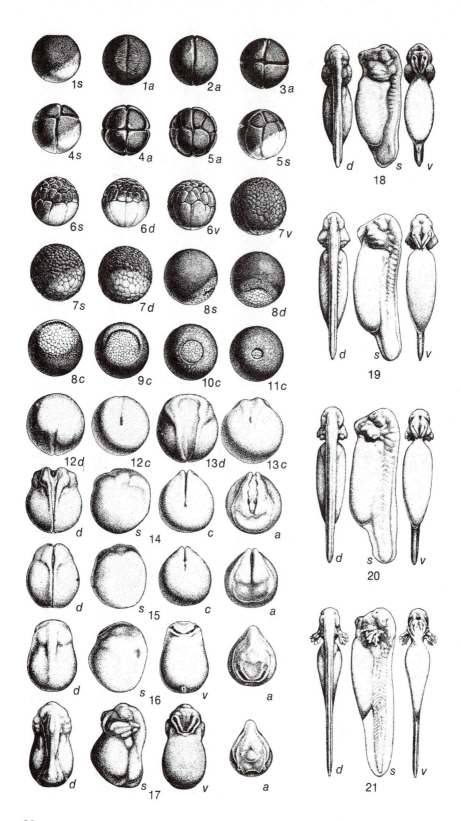

Figure 80

Stages of frog development (*Rana pipiens*). Numbers 1–33 designate developmental stages (see Table I). *a*, View from animal pole (frontal view); *c*, caudal (blastoporal) view; *d*, dorsal view; *s*, left lateral view; *v*, ventral view. Stages 1–25 6.5X; stages 26–28 mag. 3X; stages 29–33 mag. 1.2X. (From *Development of the Vertebrates* by Emil Witschi. Copyright 1956 by W. B. Saunders Co. Used with permission of the W. B. Saunders Co., a division of CBS College Publishing.

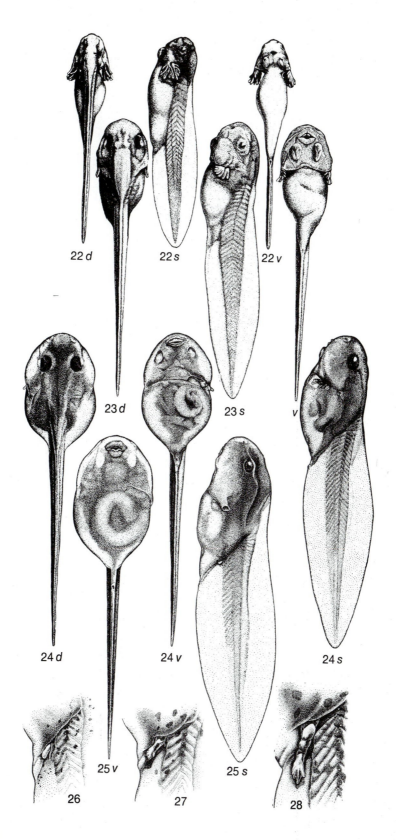

22 d 22 s 22 v

23 d 23 s v

24 d 24 v 24 s

25 v 25 s

26 27 28

Figure 81

Stages of frog development (*Rana pipiens*). Numbers 1–33 designate developmental stages (see Table I). *a*, View from animal pole (frontal view); *c*, caudal (blastoporal) view; *d*, dorsal view; *s*, left lateral view; *v*, ventral view. Stages 1–25 mag. 6.5X; stages 26–28 mag. 3X; stages 29–33 mag. 1.2X. (From *Development of the Vertebates* by Emil Witschi. Copyright 1956 by W. B. Saunders Co. Used with permission of the W. B. Saunders Co., a division of CBS College Publishing.)

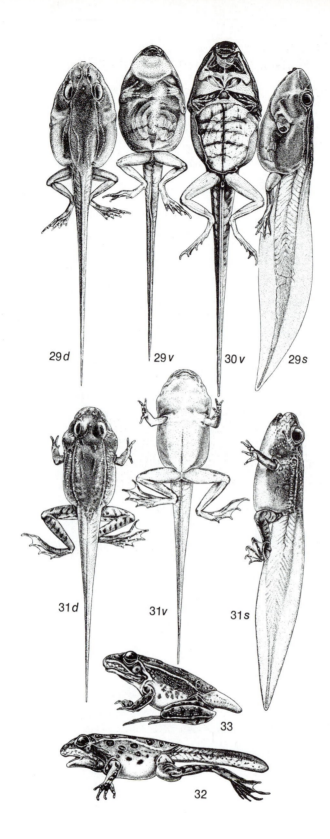

29 *d* 29 *v* 30 *v* 29 *s*

31 *d* 31 *v* 31 *s*

33

32

Figure 82

Stages of frog development (*Rana pipiens*). Numbers 1–33 designate developmental stages (see Table I). *a*, View from animal pole (frontal view); *c*, caudal (blastoporal) view; *d*, dorsal view; *s*, left lateral view; *v*, ventral view. Stages 1–25 mag. 6.5X; stages 26–28 mag. 3X; stages 29–33 mag. 1.2X. (From *Development of the Vertebates* by Emil Witschi. Copyright 1956 by W. B. Saunders Co. Used with permission of the W. B. Saunders Co., a division of CBS College Publishing.)

Table I
Frog Development Stages, *Rana pipiens* (see figs. 80–82)

Witschi Stage No.*	Approximate Lengths in mm†	Description of Stages†
1	1.7	fertilized egg
2		two-cell stage
3		four-cell stage
4		eight-cell stage
5		sixteen-cell stage
6		early blastula
7		late blastula
8		early gastrula
9		middle gastrula
10		yolk plug stage
11		late gastrula
12		neural plate stage
13		neural fold stage
14		early neural groove stage
15		late neural groove stage
16	2.5–2.7	early neural tube stage
17	2.8–3.0	early tail bud stage
18	4	tail bud stage
19	5	gill buds
20	6	hatching, gill circulation
21	7	mouth open
22	8	tail fin circulation
23	9	opercular fold
24	10	right operculum closed
25	11	operculum complete
26-33		metamorphosis

*Similar to but not identical with Shumway stage numbers.
†Based on Shumway, W., Stages in the normal development of *Rana pipiens*. I. External form. *Anatomical Record* **78**:139-147, 1940.

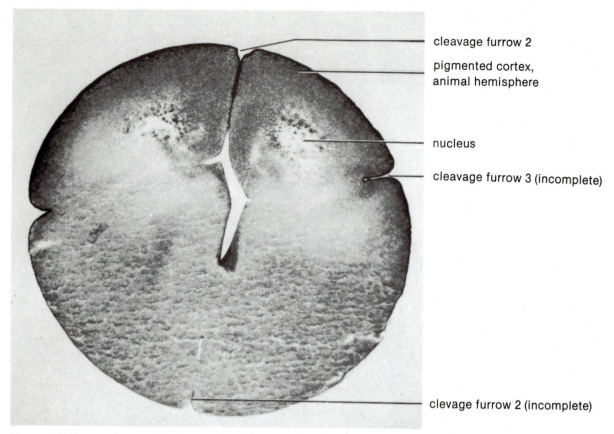

cleavage furrow 2

pigmented cortex, animal hemisphere

nucleus

cleavage furrow 3 (incomplete)

clevage furrow 2 (incomplete)

Figure 83 Frog embryo, early cleavage, 8-cell stage (Witschi stage 4), median section (mag. 65X)

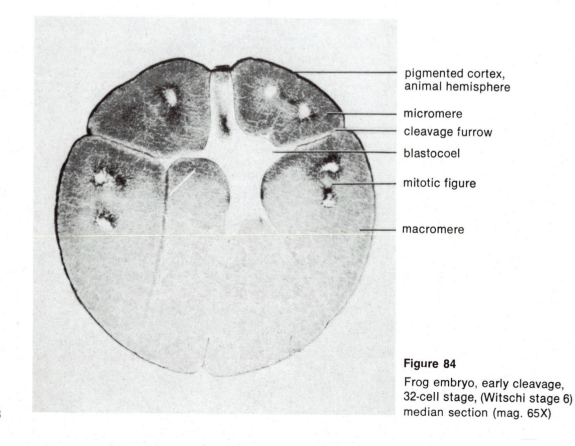

pigmented cortex, animal hemisphere

micromere

cleavage furrow

blastocoel

mitotic figure

macromere

Figure 84

Frog embryo, early cleavage, 32-cell stage, (Witschi stage 6) median section (mag. 65X)

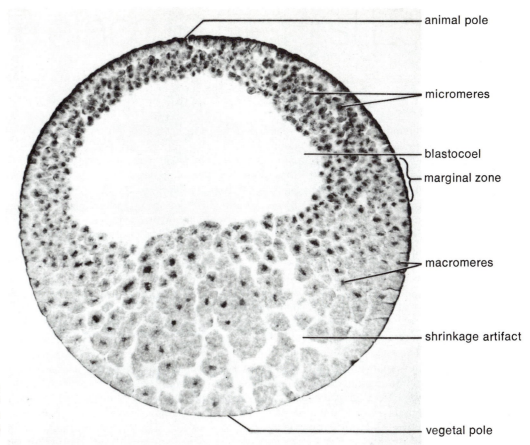

animal pole

micromeres

blastocoel

marginal zone

macromeres

shrinkage artifact

Figure 85 Frog embryo, late cleavage, blastula stage (Witschi stage 7), median section (mag. 65X)

vegetal pole

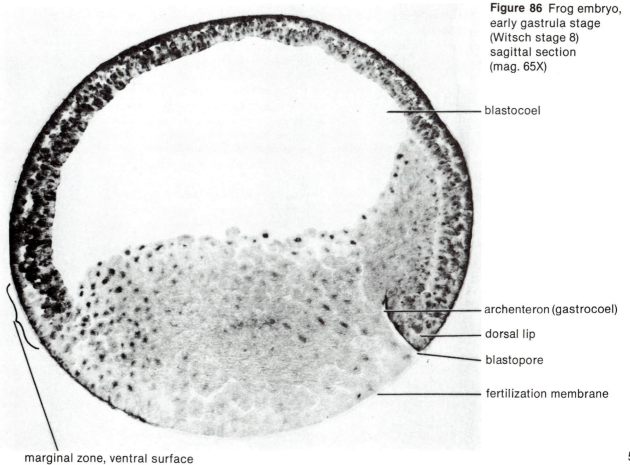

Figure 86 Frog embryo, early gastrula stage (Witsch stage 8) sagittal section (mag. 65X)

blastocoel

archenteron (gastrocoel)

dorsal lip

blastopore

fertilization membrane

marginal zone, ventral surface

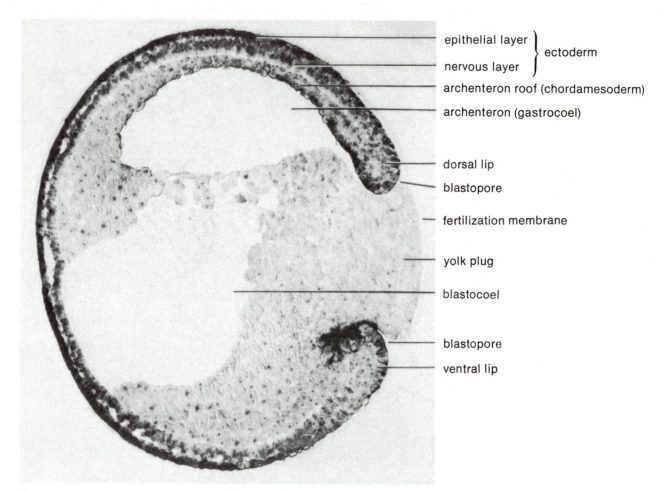

epithelial layer ⎤
　　　　　　　 ⎬ ectoderm
nervous layer ⎦

archenteron roof (chordamesoderm)

archenteron (gastrocoel)

dorsal lip

blastopore

fertilization membrane

yolk plug

blastocoel

blastopore

ventral lip

Figure 87 Frog embryo, late gastrula stage (Witschi stage 10), sagittal section (mag. 65X)

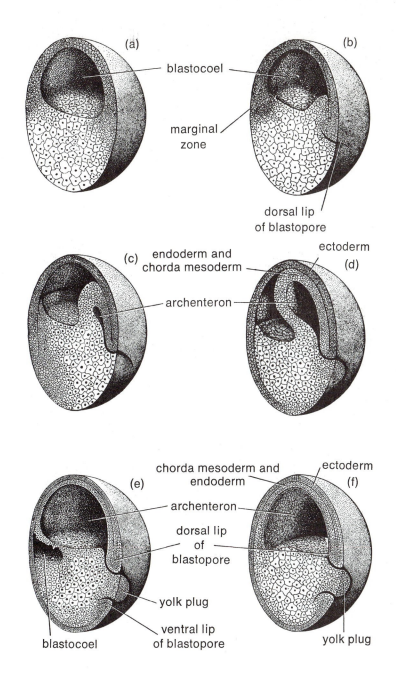

Figure 88

The blastula (*a*) of the frog and its transformation into the gastrula (*b–f*). *b*, beginning of gastrulation. *c–e*, elimination of the blastocoel or segmentation cavity by the gastrocoel or archenteron. *f*, nearly completed gastrula with mesoderm and endoderm beneath the ectoderm. (From *Fundamentals of Comparative Embryology of the Vertebrates* by Alfred F. Heuttner. Copyright 1941 by Macmillan Publishing Company, New York. Used with permissin of Macmillan Publishing Company.)

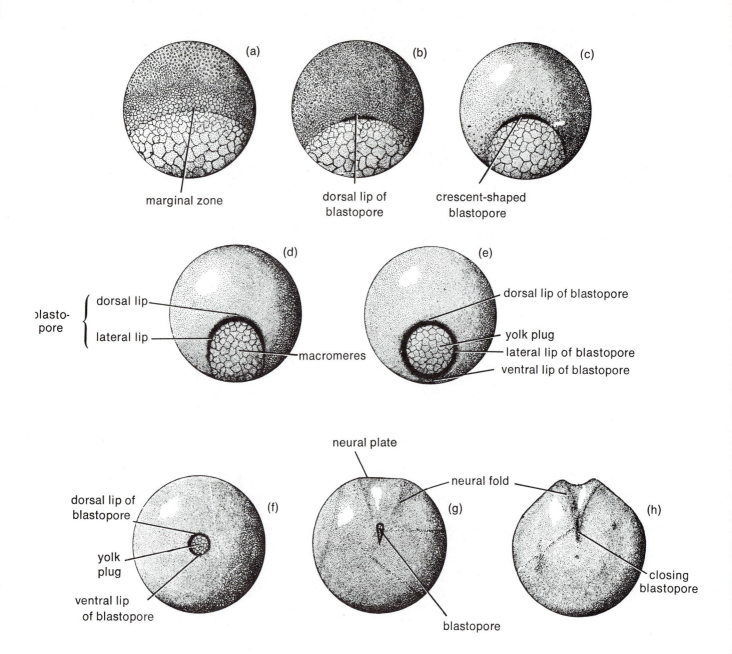

Figure 89

The process of gastrulation in the frog embryo seen from the posterior or blastoporal point of view. Stages a–e are equivalent to those contained in Fig. 88. Stages *f* and *g* are almost the same as those of *d* and *e* of Fig. 90. (From *Fundamentals of Comparative Embryology of the Vertebrates* by Alfred F. Huettner. Copyright 1941 by Macmillan Publishing Company, New York. Used with permission of Macmillan Publishing Company.)

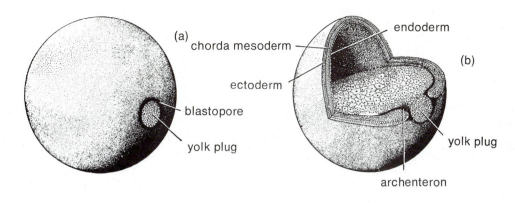

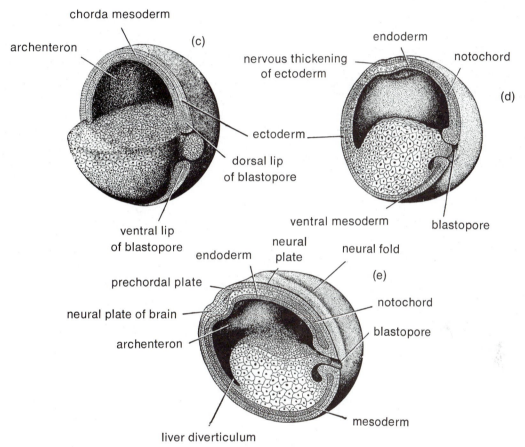

Figure 90

This figure is continuous with fig. 88. *a*, external appearance of gastrula. *b, c*, different partial sections of the gastrula. The endoderm is in the process of migrating up from the yolk cells. *d, e*, late gastrula stages. The blastopore becomes smaller, the yolk plug is withdrawn, and the embryo is elongating in the antero-posterior axis. (From *Fundamentals of Comparative Embryology of the Vertebrates* by Alfred F. Huettner. Copyright 1941 by Macmillan Publishing Company, New York. Used with permission of Macmillan Publishing Company.)

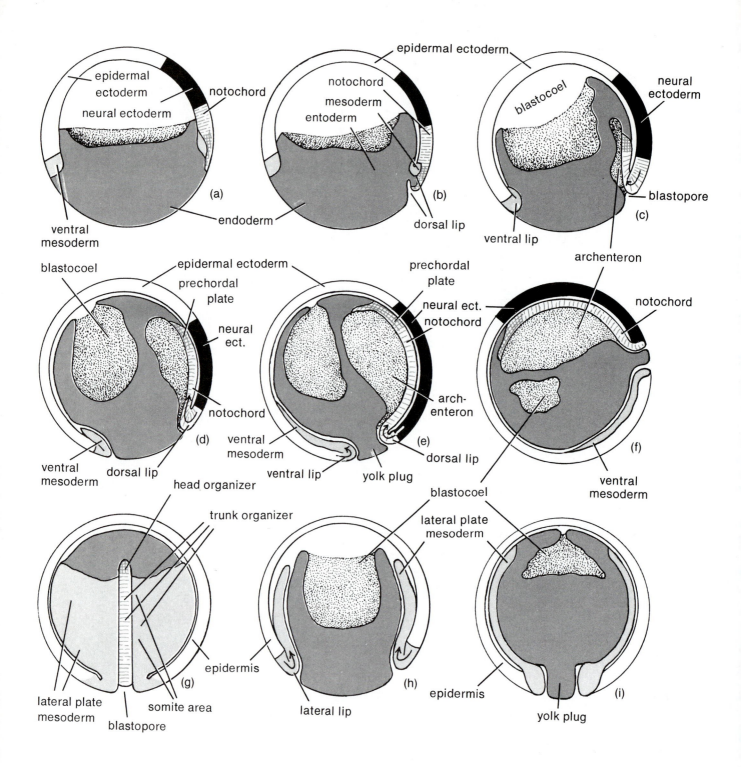

Figure 91

Migration of the presumptive organ-forming areas of the blastula during gastrulation in the amphibia (with reference particularly to the frog). *a,* late blastula, sagittal section through midplane of future embryo. *b-f,* processes of epiboly and emboly. In epiboly, the black (neural) and white (epidermal) areas become extended and gradually envelop the inward moving notochord, endoderm, and mesoderm. The processes concerned with emboly bring about the inward migration of the latter presumptive areas. *g,* late gastrular condition, with neural area and upper portion of the epidermal area removed to show relationships of the middle germ layer of chordamesoderm. *h,* horizontal section of middle gastrular condition, showing involution of mesoderm between endoderm and ectoderm. *i,* late gastrula, horizontal section, showing yolk plug, mesoderm, and final engulfment of blastocoelic space by endoderm. (From *Comparative Embryology of the Vertebrates* by Olin E. Nelson. Copyright 1953 by The Blakiston Co., Inc. Used with permission of the McGraw-Hill Book Company.)

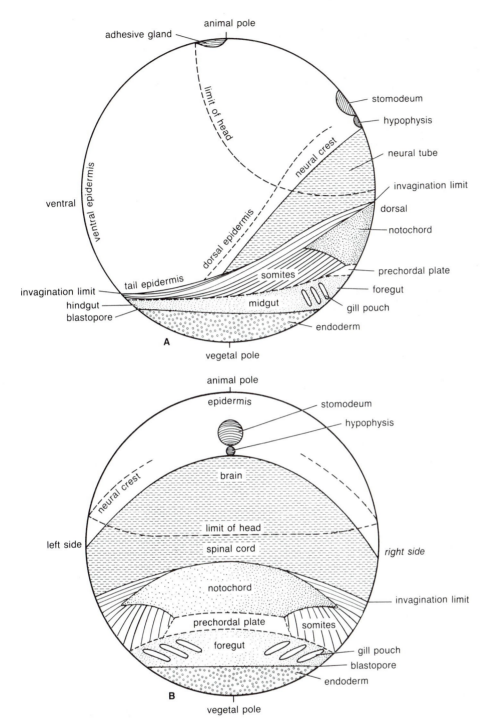

Figure 92 Maps of the presumptive regions of the very young gastrula of the anuran, *Discoglossus*, (*a*) side view. (*b*) dorsal view. [Redrawn from Jean Pasteels; New observations concerning maps of presumptive areas of the young amphibian gastrula (*Amblystoma* and *Discoglossus*). *Journal of Experimental Zoology* 89:255-251 (1943). Used with permission of the Wistar Institute Press.]

NEURAL FOLD STAGE
(Witschi Stage 14)

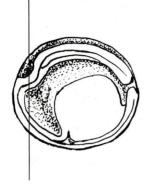

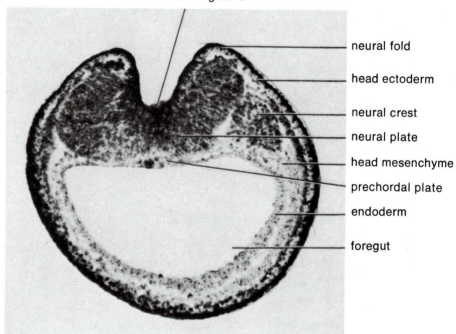

neural groove

neural fold

head ectoderm

neural crest

neural plate

head mesenchyme

prechordal plate

endoderm

foregut

Figure 93
Frog embryo, neural fold stage,
transverse section through head
region (mag. 65 ×)

66

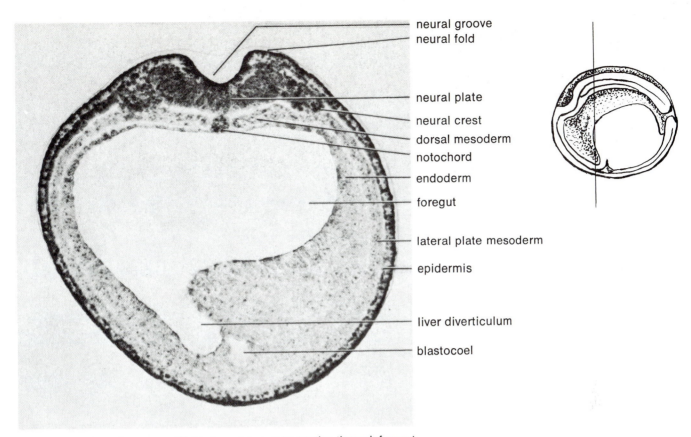

Figure 94 Frog embryo, neural fold stage, transverse section through foregut
region (mag. 65 ×)

neural groove
neural fold
neural plate
neural crest
dorsal mesoderm
notochord
endoderm
foregut
lateral plate mesoderm
epidermis
liver diverticulum
blastocoel

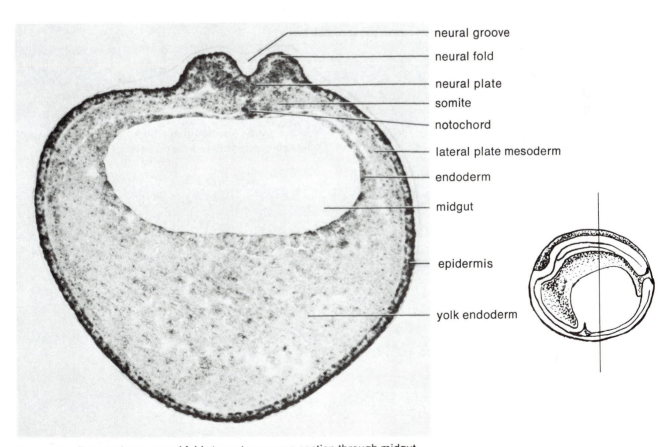

Figure 95 Frog embryo, neural fold stage, transverse section through midgut
region (mag. 65 ×)

neural groove
neural fold
neural plate
somite
notochord
lateral plate mesoderm
endoderm
midgut
epidermis
yolk endoderm

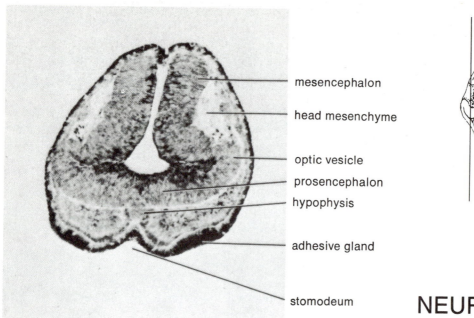

mesencephalon

head mesenchyme

optic vesicle

prosencephalon

hypophysis

adhesive gland

stomodeum

Figure 96 Frog embryo, neural tube stage, transverse section through optic vesicles (mag. 65×)

NEURAL TUBE STAGE
(Witschi Stage 16)

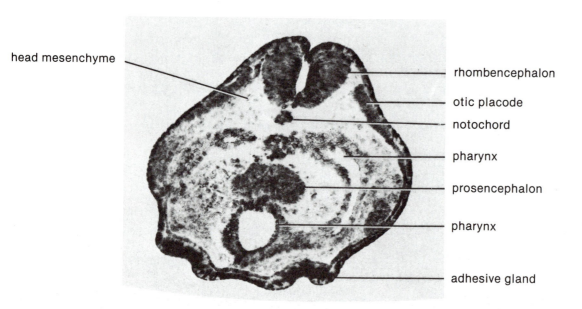

head mesenchyme

rhombencephalon

otic placode

notochord

pharynx

prosencephalon

pharynx

adhesive gland

Figure 97 Frog embryo, neural tube stage, transverse section through otic placode (mag. 65×)

96 97

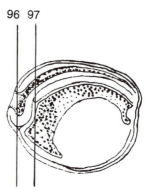

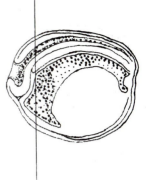

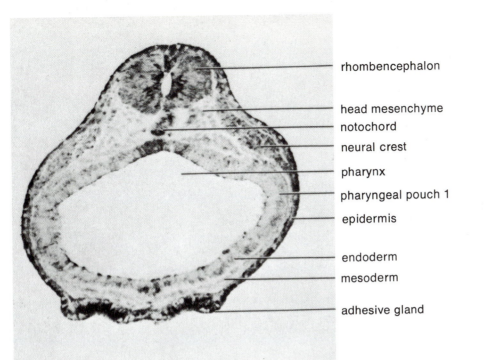

rhombencephalon

head mesenchyme
notochord
neural crest
pharynx
pharyngeal pouch 1
epidermis

endoderm
mesoderm

adhesive gland

Figure 98 Frog embryo, neural tube stage transverse
section through pharynx (mag. 65X)

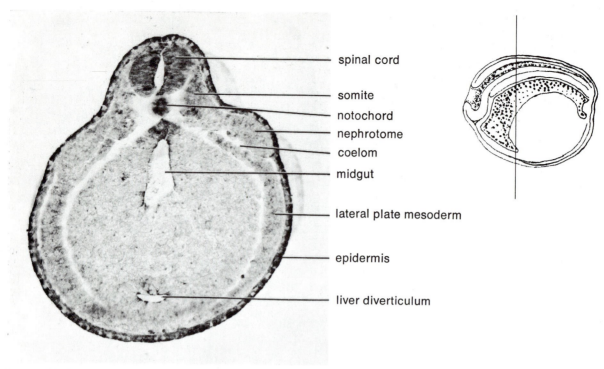

Figure 99 Frog embryo, neural tube stage transverse section through nephrotome (mag. 65X)

- spinal cord
- somite
- notochord
- nephrotome
- coelom
- midgut
- lateral plate mesoderm
- epidermis
- liver diverticulum

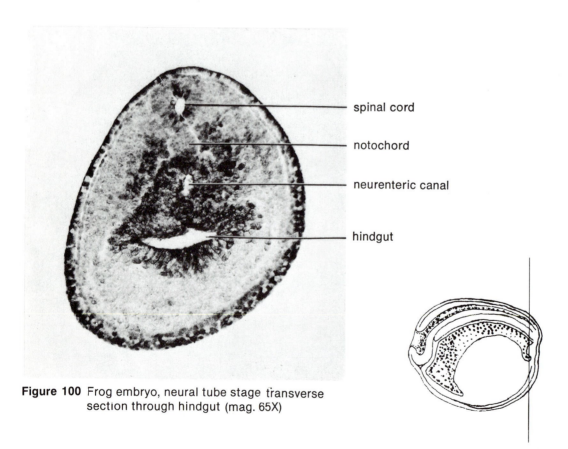

Figure 100 Frog embryo, neural tube stage transverse section through hindgut (mag. 65X)

- spinal cord
- notochord
- neurenteric canal
- hindgut

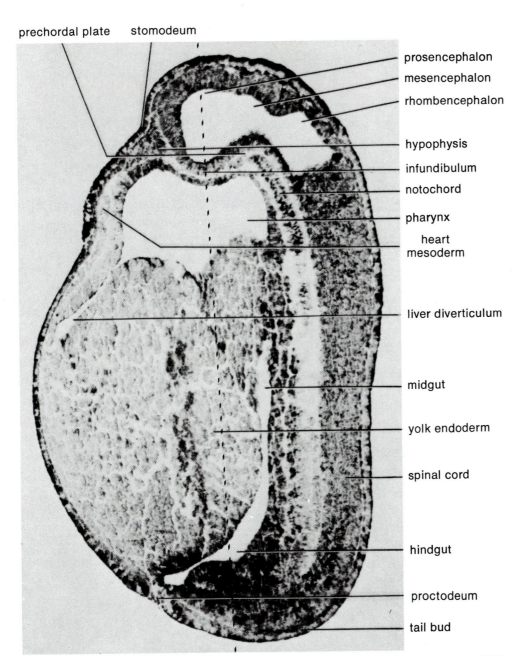

prechordal plate   stomodeum

prosencephalon

mesencephalon

rhombencephalon

hypophysis

infundibulum

notochord

pharynx

heart
mesoderm

liver diverticulum

midgut

yolk endoderm

spinal cord

hindgut

proctodeum

tail bud

Figure 101 Frog embryo, tail bud stage (Witchi stage 17), sagittal section (mag. 65X)
The dotted line indicates plane of section in fig. 102.

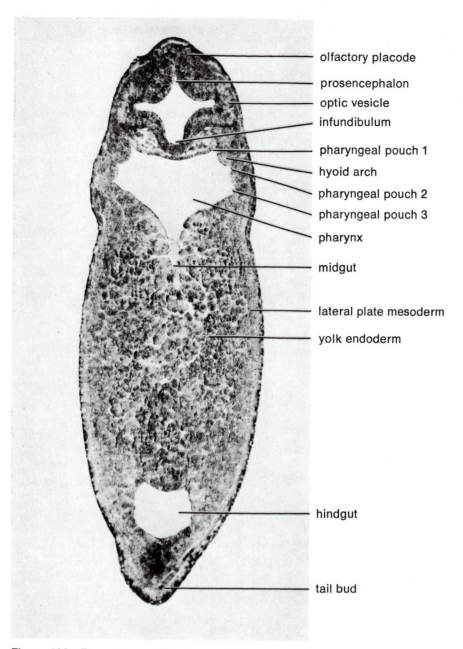

olfactory placode

prosencephalon

optic vesicle

infundibulum

pharyngeal pouch 1

hyoid arch

pharyngeal pouch 2

pharyngeal pouch 3

pharynx

midgut

lateral plate mesoderm

yolk endoderm

hindgut

tail bud

Figure 102   Frog embryo, tail bud stage (Witschi stage 17), frontal section through parynx (mag. 50X). (See fig. 101 for the plane of section.)

7. The 4-mm Frog Embryo (Stage 18)

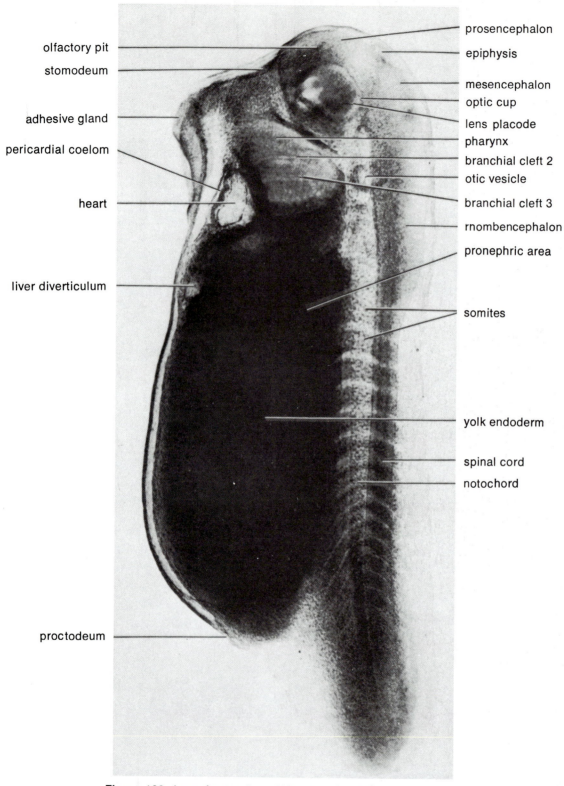

olfactory pit

stomodeum

adhesive gland

pericardial coelom

heart

liver diverticulum

proctodeum

prosencephalon

epiphysis

mesencephalon

optic cup

lens placode

pharynx

branchial cleft 2

otic vesicle

branchial cleft 3

rnombencephalon

pronephric area

somites

yolk endoderm

spinal cord

notochord

Figure 103 4-mm frog embryo (Witschi stage 18), whole mount (mag. 50X)

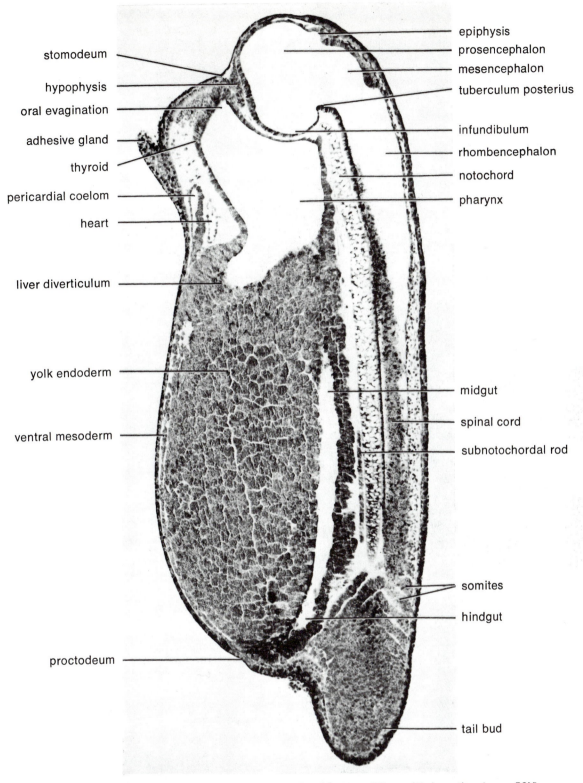

stomodeum

hypophysis

oral evagination

adhesive gland

thyroid

pericardial coelom

heart

liver diverticulum

yolk endoderm

ventral mesoderm

proctodeum

epiphysis

prosencephalon

mesencephalon

tuberculum posterius

infundibulum

rhombencephalon

notochord

pharynx

midgut

spinal cord

subnotochordal rod

somites

hindgut

tail bud

Figure 104 4-mm frog embryo (Witschi stage 18), sagittal section (mag. 50X)

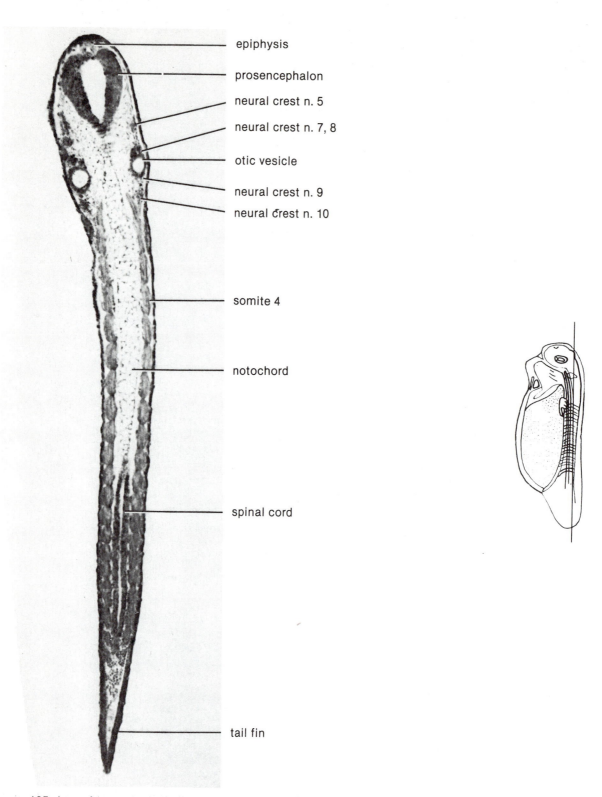

epiphysis

prosencephalon

neural crest n. 5

neural crest n. 7, 8

otic vesicle

neural crest n. 9

neural crest n. 10

somite 4

notochord

spinal cord

tail fin

Figure 105 4-mm frog embryo (Witschi stage 18), frontal section through otic vesicles (mag. 40X)

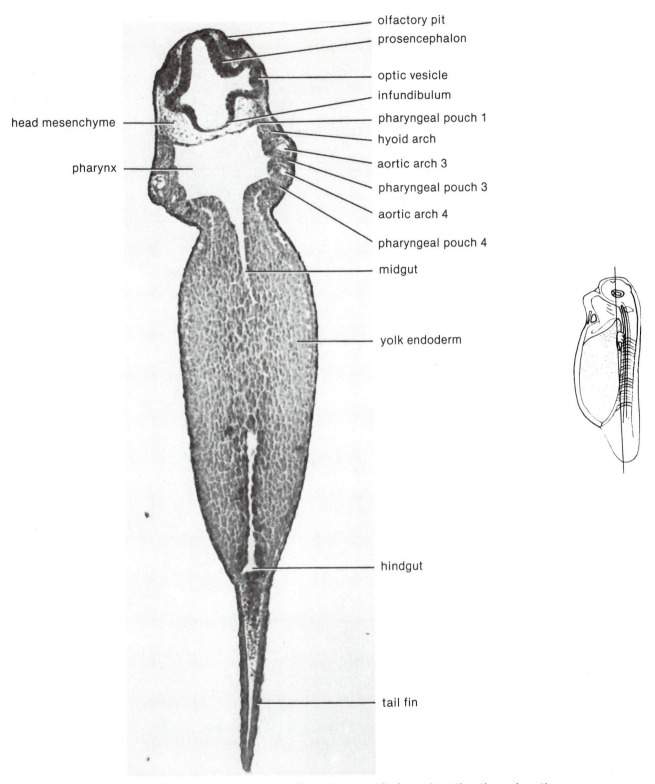

olfactory pit
prosencephalon
optic vesicle
infundibulum
pharyngeal pouch 1
hyoid arch
aortic arch 3
pharyngeal pouch 3
aortic arch 4
pharyngeal pouch 4
midgut
yolk endoderm
hindgut
tail fin
head mesenchyme
pharynx

Figure 106 4-mm frog embryo (Witschi stage 18), frontal section through optic
vesicles (mag. 40X)

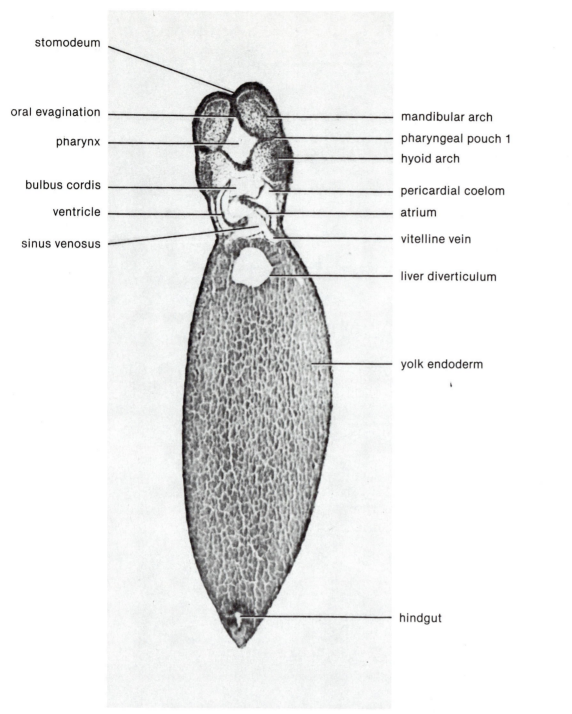

stomodeum

oral evagination

pharynx

bulbus cordis

ventricle

sinus venosus

mandibular arch

pharyngeal pouch 1

hyoid arch

pericardial coelom

atrium

vitelline vein

liver diverticulum

yolk endoderm

hindgut

Figure 107 4-mm frog embryo (Witschi stage 18), frontal section through heart (mag. 40X)

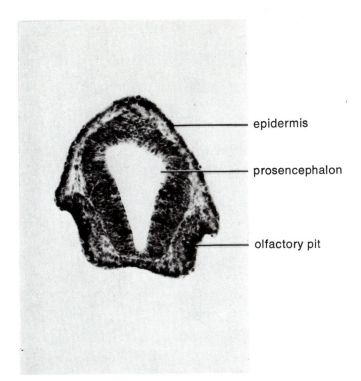

Figure 108 4-mm frog embryo (Witschi stage 18), transverse section through olfactory pits (mag. 65X)

epidermis

prosencephalon

olfactory pit

Figure 109 4-mm frog embryo, transverse section through optic cups (mag. 65X)

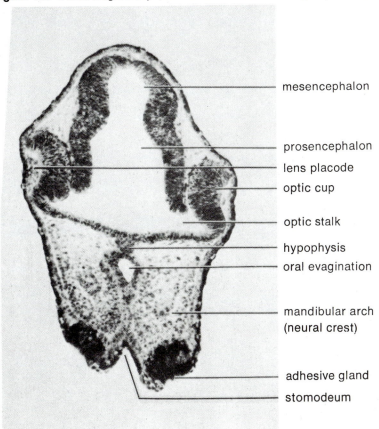

mesencephalon

prosencephalon

lens placode

optic cup

optic stalk

hypophysis

oral evagination

mandibular arch
(neural crest)

adhesive gland

stomodeum

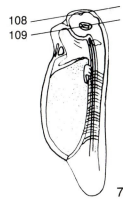

108
109

79

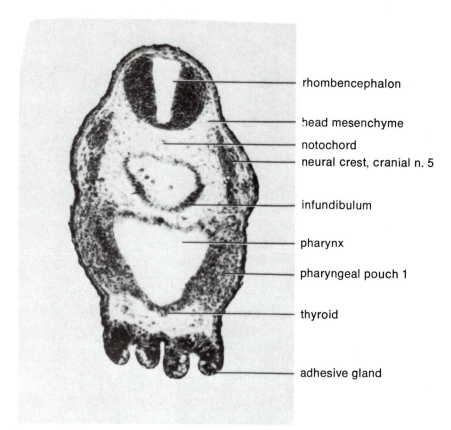

Figure 110 4-mm frog embryo, transverse section through anterior pharynx (mag. 65X)

rhombencephalon

head mesenchyme

notochord

neural crest, cranial n. 5

infundibulum

pharynx

pharyngeal pouch 1

thyroid

adhesive gland

Figure 111 4-mm frog embryo, transverse section through otic vesicles (mag. 65X)

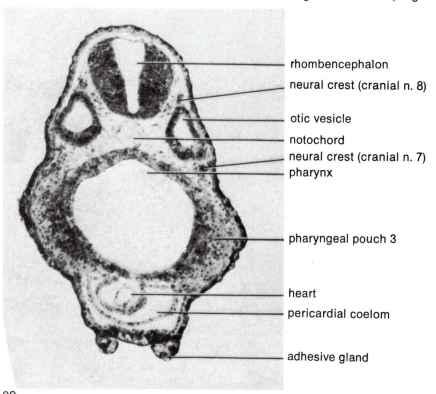

rhombencephalon

neural crest (cranial n. 8)

otic vesicle

notochord

neural crest (cranial n. 7)

pharynx

pharyngeal pouch 3

heart

pericardial coelom

adhesive gland

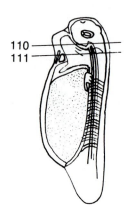

110

111

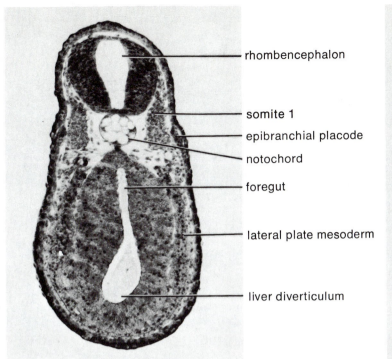

rhombencephalon

somite 1

epibranchial placode

notochord

foregut

lateral plate mesoderm

liver diverticulum

Figure 112 4-mm frog embryo, transverse section through liver diverticulum (mag. 65X)

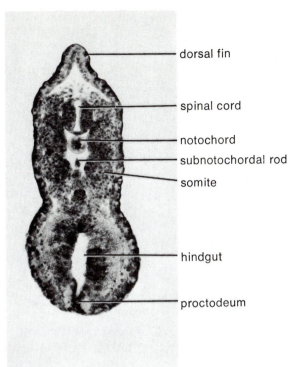

dorsal fin

spinal cord

notochord

subnotochordal rod

somite

hindgut

proctodeum

Figure 114 4-mm frog embryo, transverse section through hindgut (mag. 65X)

Figure 113 4-mm frog embryo, transverse section through pronephros (mag. 65X)

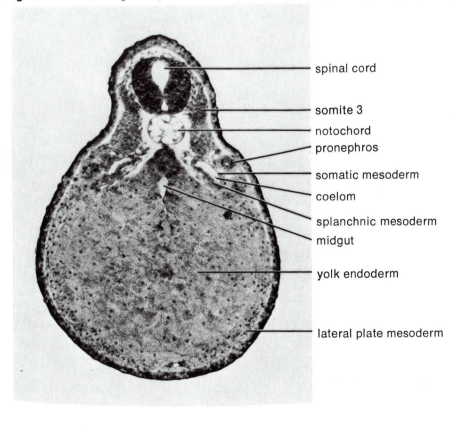

spinal cord

somite 3

notochord

pronephros

somatic mesoderm

coelom

splanchnic mesoderm

midgut

yolk endoderm

lateral plate mesoderm

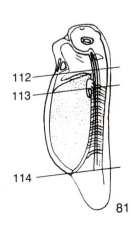

112

113

114

8. The 6–7-mm Frog Tadpole (Stages 20–21)

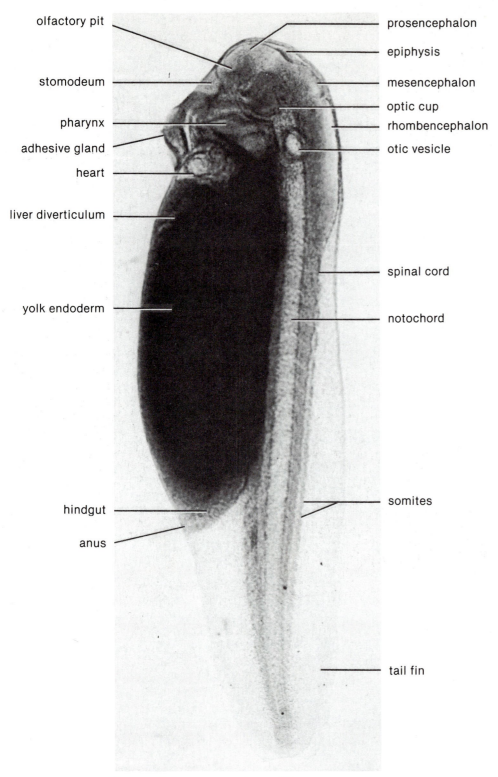

olfactory pit

stomodeum

pharynx

adhesive gland

heart

liver diverticulum

yolk endoderm

hindgut

anus

prosencephalon

epiphysis

mesencephalon

optic cup

rhombencephalon

otic vesicle

spinal cord

notochord

somites

tail fin

Figure 115 6-mm frog tadpole (Witschi stage 20),
whole mount (mag. 35X)

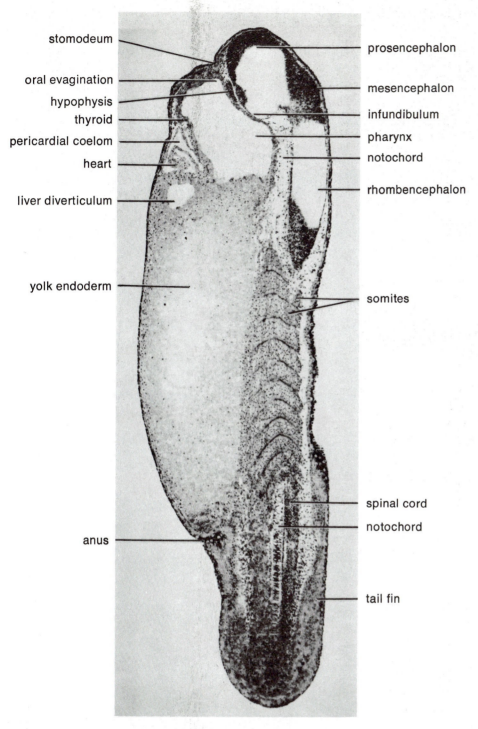

stomodeum

oral evagination

hypophysis

thyroid

pericardial coelom

heart

liver diverticulum

yolk endoderm

anus

prosencephalon

mesencephalon

infundibulum

pharynx

notochord

rhombencephalon

somites

spinal cord

notochord

tail fin

Figure 116 6-mm frog tadpole, sagittal section (Witschi stage 20), (mag. 35×)

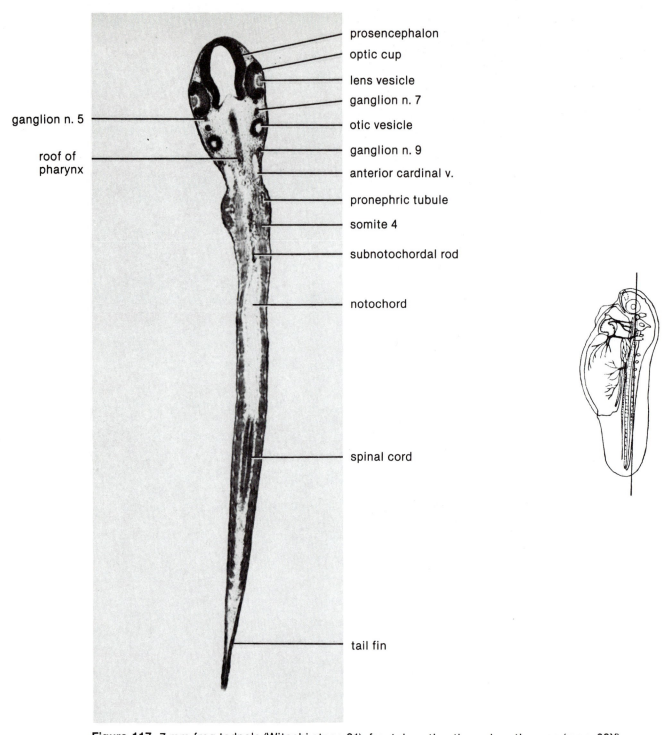

prosencephalon
optic cup
lens vesicle
ganglion n. 7
ganglion n. 5
otic vesicle
roof of
pharynx
ganglion n. 9
anterior cardinal v.
pronephric tubule
somite 4
subnotochordal rod

notochord

spinal cord

tail fin

Figure 117 7-mm frog tadpole (Witschi stage 21), frontal section through optic cups (mag. 30X) ·

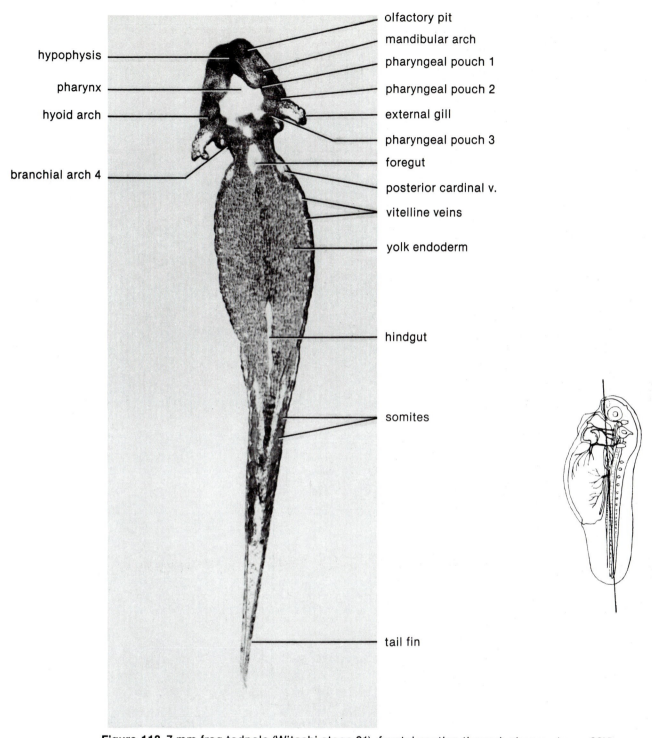

hypophysis

pharynx

hyoid arch

branchial arch 4

olfactory pit

mandibular arch

pharyngeal pouch 1

pharyngeal pouch 2

external gill

pharyngeal pouch 3

foregut

posterior cardinal v.

vitelline veins

yolk endoderm

hindgut

somites

tail fin

Figure 118 7-mm frog tadpole (Witschi stage 21), frontal section through pharynx (mag. 30X)

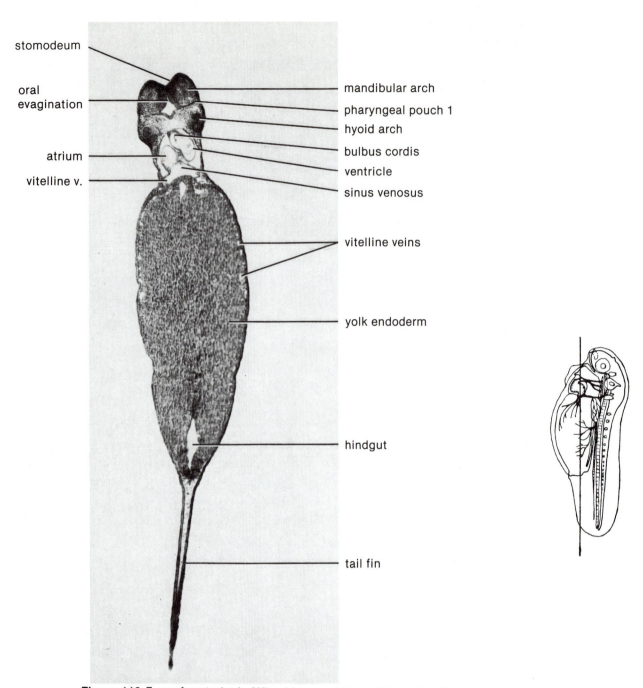

stomodeum

oral
evagination

atrium

vitelline v.

mandibular arch

pharyngeal pouch 1

hyoid arch

bulbus cordis

ventricle

sinus venosus

vitelline veins

yolk endoderm

hindgut

tail fin

Figure 119 7-mm frog tadpole (Witschi stage 21), frontal section through heart (mag. 30X)

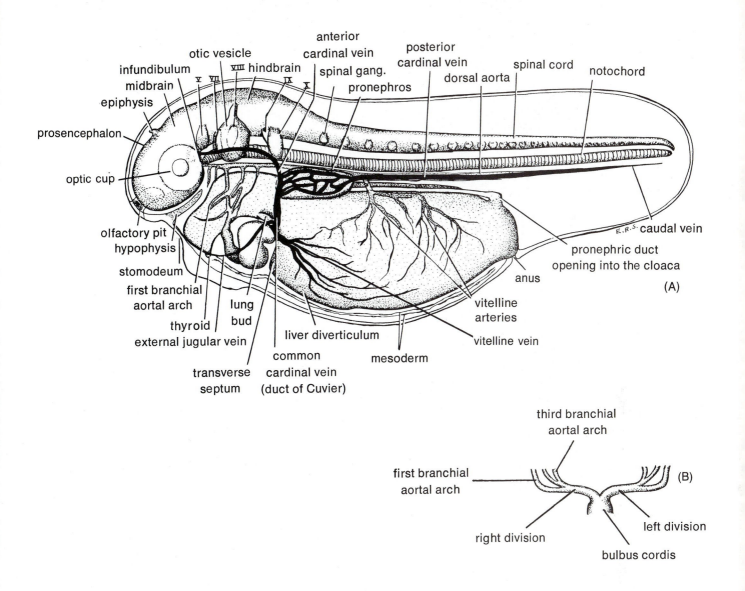

Figure 120 (A) Drawing of 6–7-mm frog tadpole (From *Comparative Embryology of the Vertebrates* by Olin E. Nelson. Copyright 1953 by The Blakiston Co. Inc. Used with permission of McGraw-Hill Book Company. (B) Drawing of the aortic arches of 6–7 mm frog tadpole.

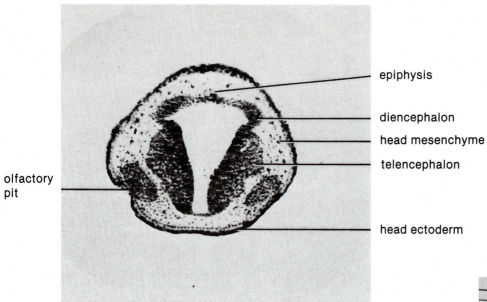

epiphysis

diencephalon

head mesenchyme

telencephalon

olfactory pit

head ectoderm

Figure 121 7-mm frog tadpole (Witschi stage 21), transverse section through olfactory pit (mag. 65X)

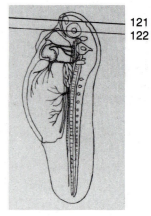

121
122

head mesenchyme

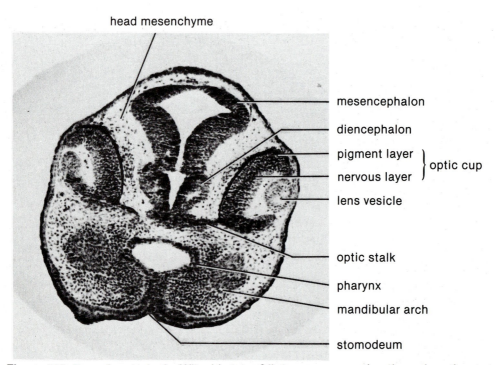

mesencephalon

diencephalon

pigment layer ⎫
⎬ optic cup
nervous layer ⎭

lens vesicle

optic stalk

pharynx

mandibular arch

stomodeum

Figure 122 7-mm frog tadpole (Witschi stage 21), transverse section through optic cups (mag. 65X)

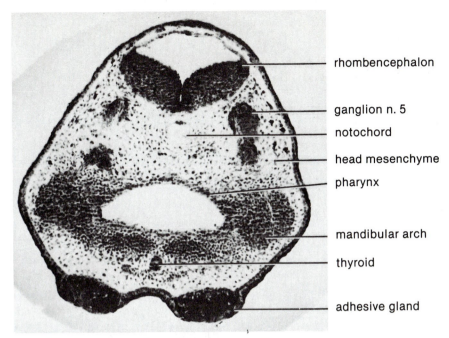

— rhombencephalon

— ganglion n. 5
— notochord

— head mesenchyme

— pharynx

— mandibular arch

— thyroid

— adhesive gland

Figure 123 7-mm frog tadpole (Witschi stage 21), transverse section through thyroid (mag. 65X)

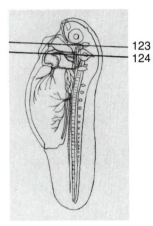

123
124

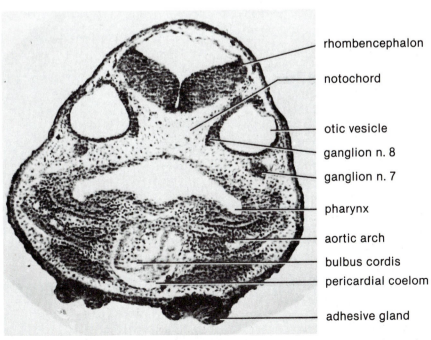

— rhombencephalon

— notochord

— otic vesicle

— ganglion n. 8

— ganglion n. 7

— pharynx

— aortic arch

— bulbus cordis

— pericardial coelom

— adhesive gland

Figure 124 7-mm frog tadpole (Witschi stage 21), transverse section through otic vesicle (mag. 65X)

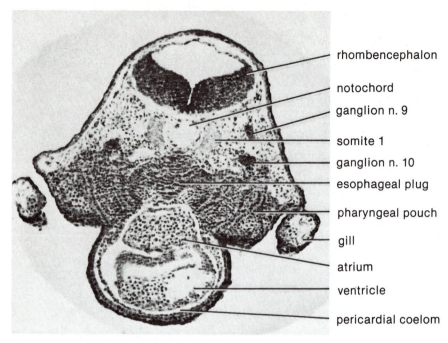

rhombencephalon

notochord

ganglion n. 9

somite 1

ganglion n. 10

esophageal plug

pharyngeal pouch

gill

atrium

ventricle

pericardial coelom

Figure 125 7-mm frog tadpole (Witschi stage 21), transverse section through heart (mag. 65X)

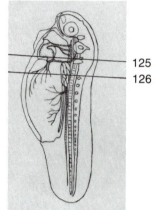

125
126

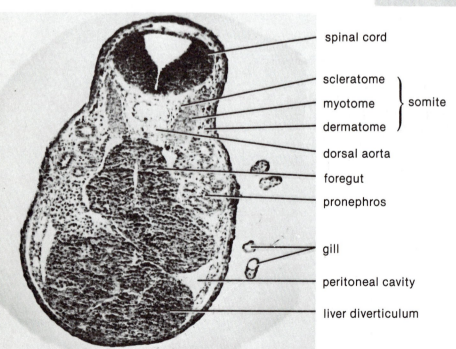

spinal cord

scleratome ⎫
myotome   ⎬ somite
dermatome ⎭

dorsal aorta

foregut

pronephros

gill

peritoneal cavity

liver diverticulum

Figure 126 7-mm frog tadpole (Witschi stage 21), transverse section through pronephrous (mag. 65X)

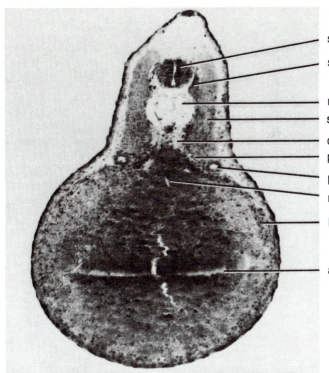

spinal cord

spinal ganglion

notochord

somite

dorsal aorta

posterior cardinal v.

pronephric duct

midgut

lateral plate

artifact

Figure 127 7-mm frog tadpole (Witschi stage 21), transverse section through midgut (mag. 65X)

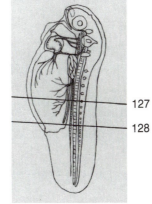

127

128

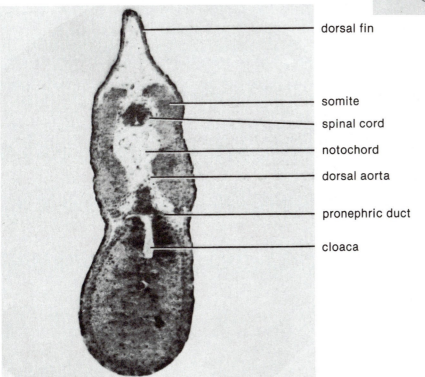

dorsal fin

somite

spinal cord

notochord

dorsal aorta

pronephric duct

cloaca

Figure 128 7-mm frog tadpole (Witschi stage 21), transverse section through cloaca (mag. 65X)

93

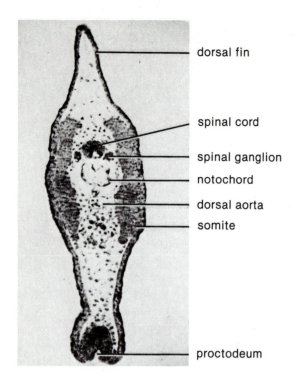

dorsal fin

spinal cord

spinal ganglion

notochord

dorsal aorta

somite

proctodeum

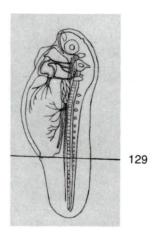

129

Figure 129 7-mm frog tadpole (Witschi stage 21), transverse section through proctodeum (mag. 65X)

9. The 10–mm Frog Tadpole (Stage 24)

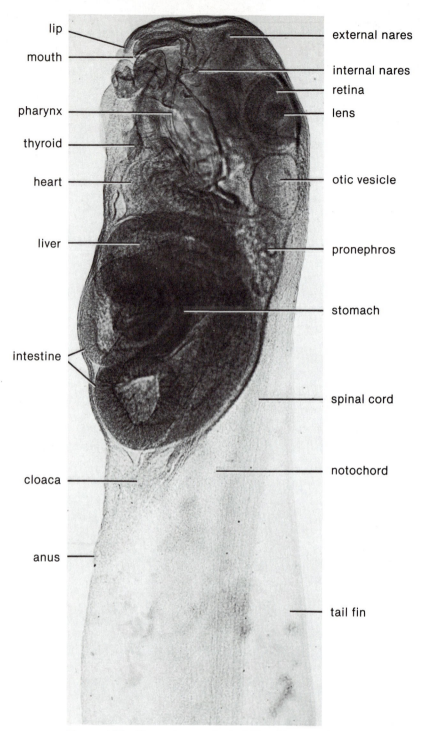

lip

mouth

pharynx

thyroid

heart

liver

intestine

cloaca

anus

external nares

internal nares

retina

lens

otic vesicle

pronephros

stomach

spinal cord

notochord

tail fin

Figure 130 10-mm frog tadpole (Witschi stage 24), whole mount (mag. 40X)

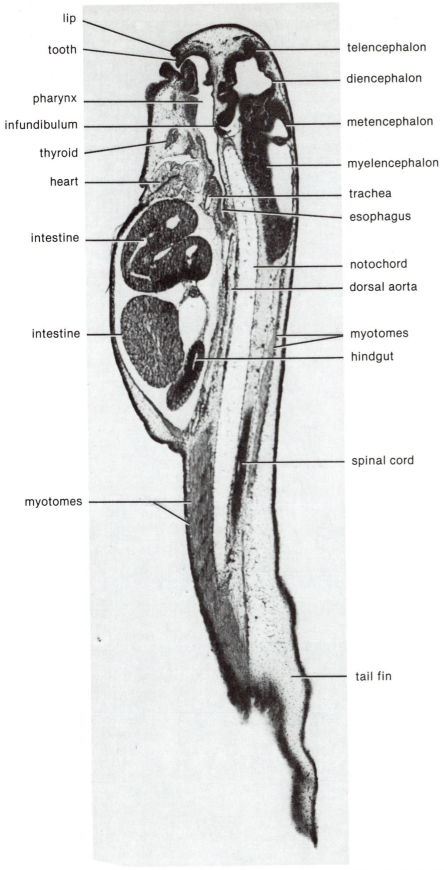

lip

tooth

pharynx

infundibulum

thyroid

heart

intestine

intestine

myotomes

telencephalon

diencephalon

metencephalon

myelencephalon

trachea

esophagus

notochord

dorsal aorta

myotomes

hindgut

spinal cord

tail fin

Figure 131 10-mm frog tadpole (Witschi stage 24), sagittal section (mag. 30X)

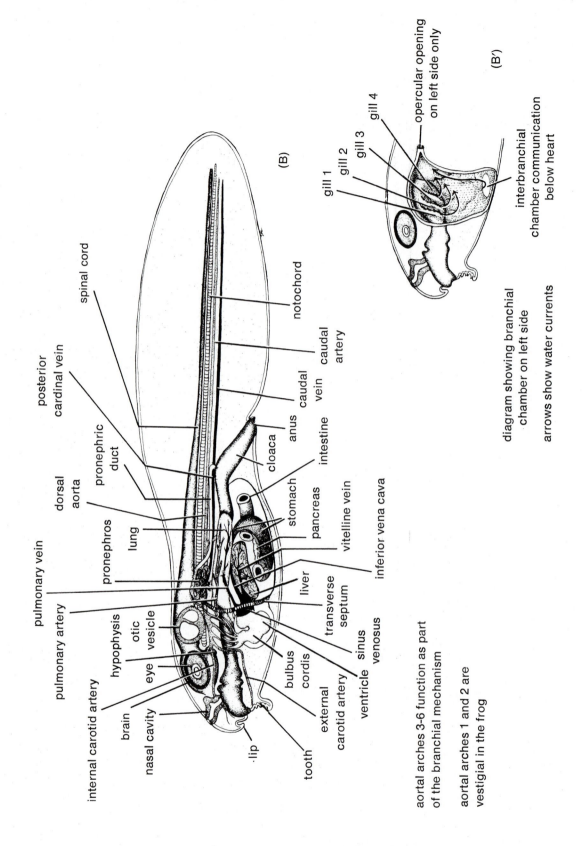

spinal cord

posterior
cardinal vein

pronephric
duct

dorsal
aorta

pronephros

lung

pulmonary vein

pulmonary artery

hypophysis

otic
vesicle

internal carotid artery

brain

nasal cavity

eye

lip

tooth

external
carotid artery

bulbus
cordis

ventricle

sinus
venosus

transverse
septum

liver

inferior vena cava

vitelline vein

pancreas

stomach

intestine

cloaca

anus

caudal
vein

caudal
artery

notochord

(B)

aortal arches 3-6 function as part
of the branchial mechanism

aortal arches 1 and 2 are
vestigial in the frog

gill 1

gill 2

gill 3

gill 4

opercular opening
on left side only

(B')

interbranchial
chamber communication
below heart

diagram showing branchial
chamber on left side

arrows show water currents

Figure 132 Drawing of 10–18-mm frog tadpole. (From *Comparative Embryology of the Vertebrates* by Olin E. Nelson. Copyright 1953 by the Blakiston Co. Inc. Used with permission of McGraw-Hill Book Company.)

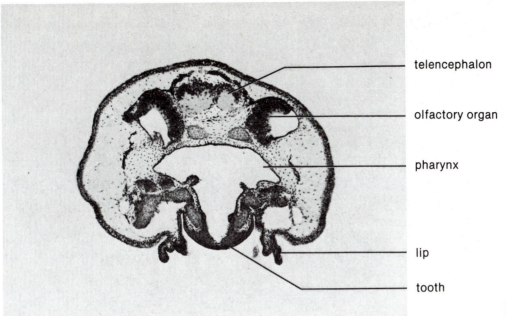

telencephalon

olfactory organ

pharynx

lip

tooth

Figure 133 10-mm frog tadpole (Witschi stage 24), transverse section through olfactory organ (mag. 50X)

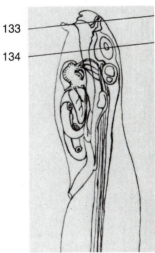

133

134

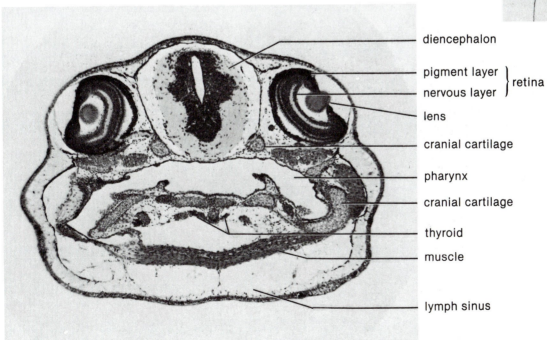

diencephalon

pigment layer
}
nervous layer } retina

lens

cranial cartilage

pharynx

cranial cartilage

thyroid

muscle

lymph sinus

Figure 134 10-mm frog tadpole (Witschi stage 24), transverse section through eyes (mag. 50X)

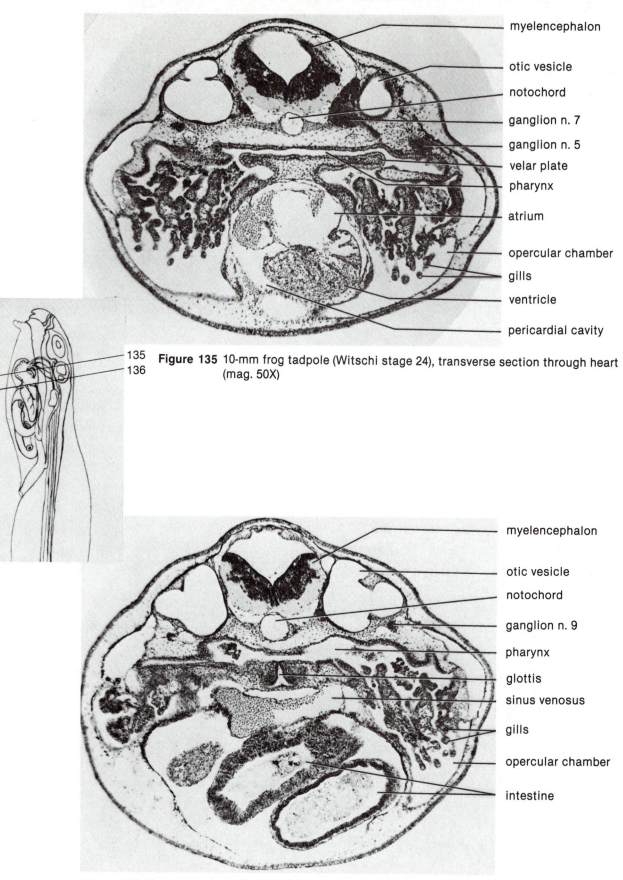

myelencephalon

otic vesicle

notochord

ganglion n. 7

ganglion n. 5

velar plate

pharynx

atrium

opercular chamber

gills

ventricle

pericardial cavity

135
136

Figure 135 10-mm frog tadpole (Witschi stage 24), transverse section through heart (mag. 50X)

myelencephalon

otic vesicle

notochord

ganglion n. 9

pharynx

glottis

sinus venosus

gills

opercular chamber

intestine

Figure 136 10-mm frog tadpole (Witschi stage 24), transverse section through glottis (mag. 50X)

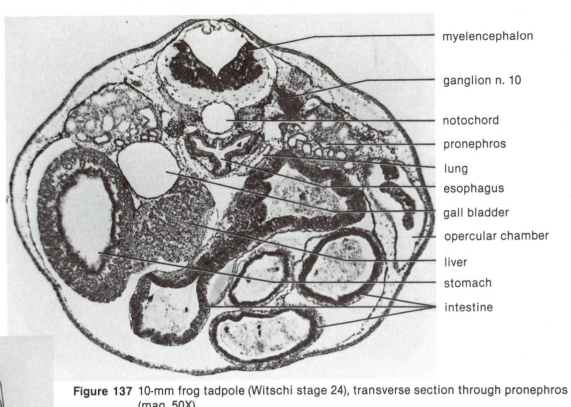

myelencephalon

ganglion n. 10

notochord

pronephros

lung

esophagus

gall bladder

opercular chamber

liver

stomach

intestine

137
138

Figure 137 10-mm frog tadpole (Witschi stage 24), transverse section through pronephros (mag. 50X)

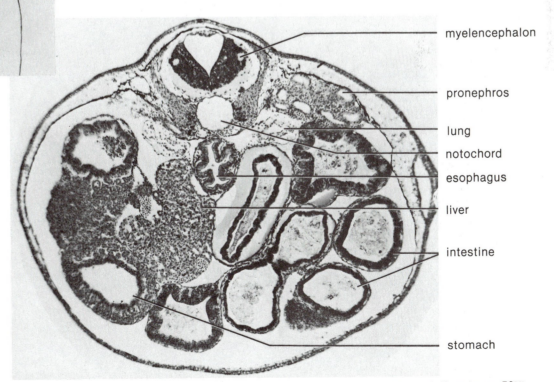

myelencephalon

pronephros

lung

notochord

esophagus

liver

intestine

stomach

Figure 138 10-mm frog tadpole (Witschi stage 24), transverse section through liver (mag. 50X)

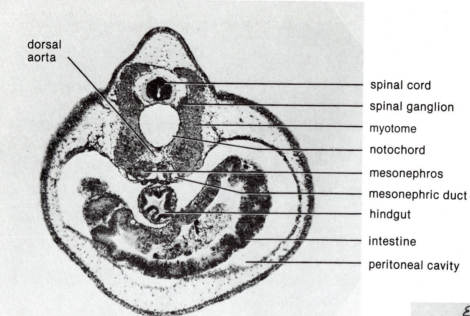

dorsal aorta

spinal cord
spinal ganglion
myotome
notochord
mesonephros
mesonephric duct
hindgut
intestine
peritoneal cavity

Figure 139 10-mm frog tadpole (Witschi stage 24), transverse section through mesonephros (mag. 50X)

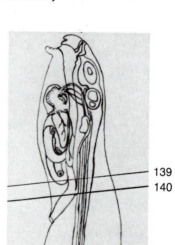

139
140

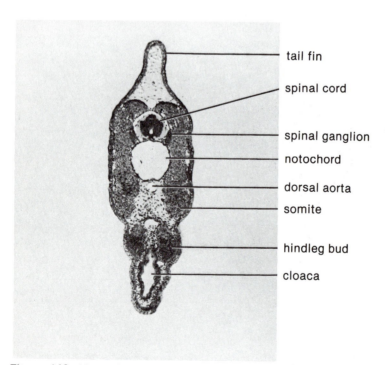

tail fin

spinal cord

spinal ganglion
notochord
dorsal aorta
somite

hindleg bud
cloaca

Figure 140 10-mm frog tadpole (Witschi stage 24), transverse section through cloaca (mag. 50X)

102

10. Gametogenesis in Chickens

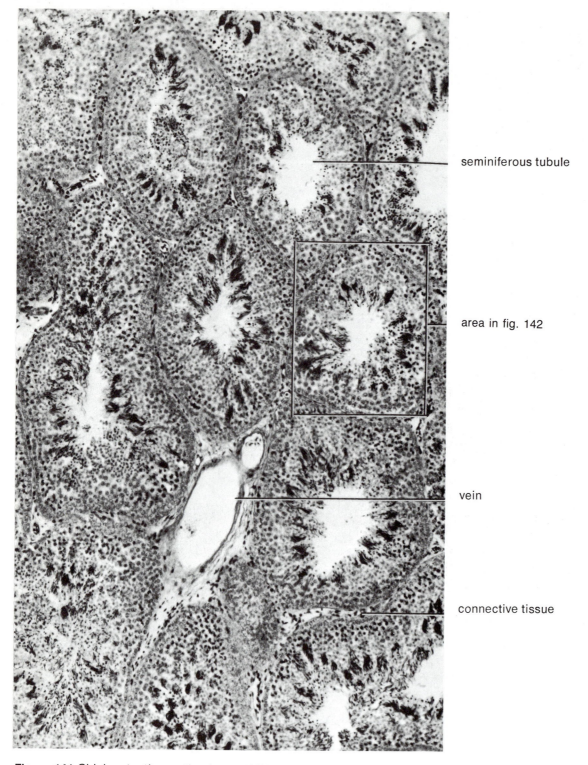

seminiferous tubule

area in fig. 142

vein

connective tissue

Figure 141 Chicken testis, section (mag. 180X)

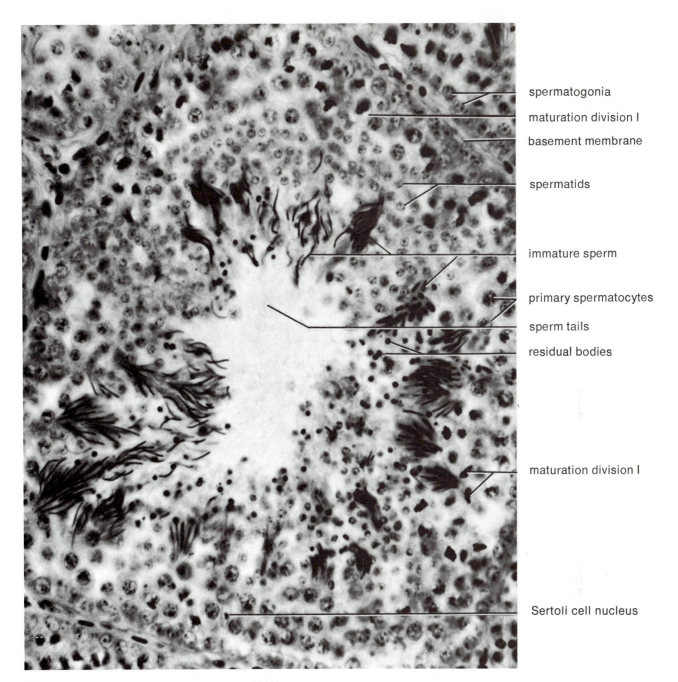

spermatogonia

maturation division I

basement membrane

spermatids

immature sperm

primary spermatocytes

sperm tails

residual bodies

maturation division I

Sertoli cell nucleus

Figure 142 Chicken testis, section (mag. 680X)

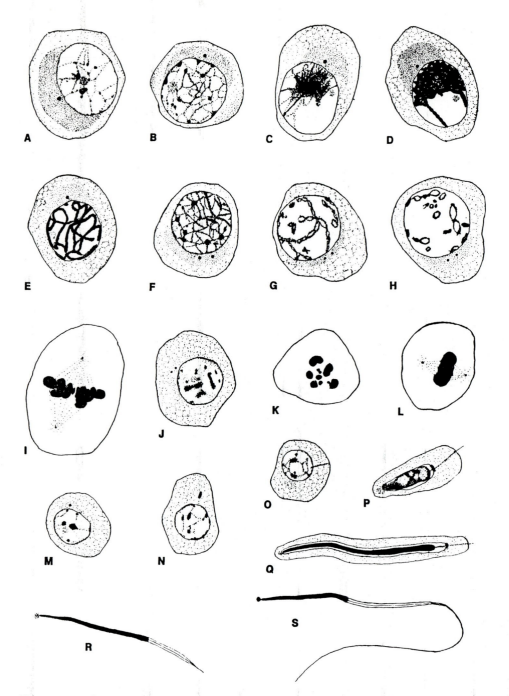

Figure 143 Spermatogenesis in the chicken. *a–i*, primary spermatocyte. *j–l*, secondary spermatocyte. *m*, spermatid. *n–s*, spermeogenesis. *a*, interphase. *b*, leptotene. *c*, zygotene. *d, e*, pachytene. *f*, diffuse stage. *g, h*, diplotene. *i*, metaphase first maturation division. *j*, interphase of secondary spermtocyte. *k, l*, metaphase second maturation division. (From R. A. Miller, Spermatogeneis in a sex-reversed female and in normal males of the domestic fowl, *Gallus domesticus. Anatomical Record*, 1938. Used with permission of the Wistar Institute Press.)

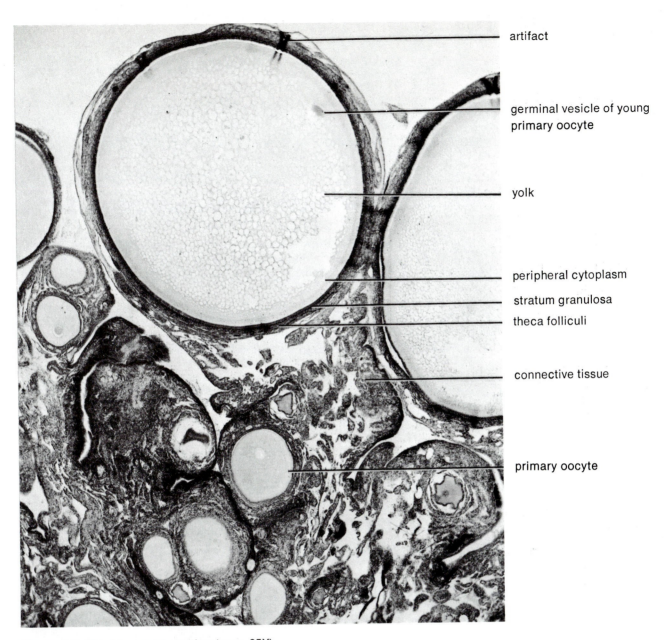

artifact

germinal vesicle of young
primary oocyte

yolk

peripheral cytoplasm

stratum granulosa

theca folliculi

connective tissue

primary oocyte

Figure 144 Chicken ovary, section (mag. 35X)

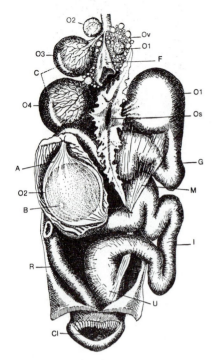

Figure 145 The female reproductive system of the fowl (After Coste-Duval.) The figure shows two eggs in the oviduct, though normally only one is in the oviduct at a time. A, albumen (dense layer); B, blastoderm; C, cicatrix; cl, cloaca; F, follicle from which egg has been discharged; G, Glandular portion of oviduct, I, isthmus; O_1–O_4, ovarian ova in different stages of growth, each enclosed in a follicle richly supplied with blood vessels; O_1, ovum in upper end of oviduct; O_2, ovum in middle portion of oviduct, which has been cut open to show the ovum in position; Os, ostium (infundibulum) of oviduct; Ov, ovary, R, rectum, U, uterus. (From H. L. Wieman, *An Introduction to Vertebrate Embryology*. Copyright 1949 by McGraw-Hill Book Co. Used with permission of McGraw-Hill Book Co.)

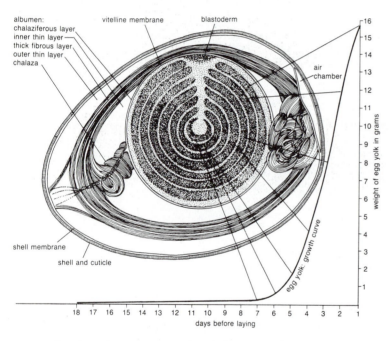

Figure 146 Growth and composition of hen's egg before incubation. (Weight curve of ovocyte during 18 days preceding laying, after Gerhartz 1914.) The cells of the blastoderm are drawn relatively too large in size and too few in number. White yolk is lightly strippled, yellow yolk is heavily stippled (the central white yolk is called latebra of Purkinje, that below the blastoderm nucleus of Pander). Egg yolk, being relatively lighter than egg white, tends to float toward the highest part of the shell. (From *Development of the Vertebrates*, by Emil Witschi. Copyright © 1956 by W. B. Saunders Company. Used with permission of the W. B. Saunders Company, a division of CBS College Publishing.)

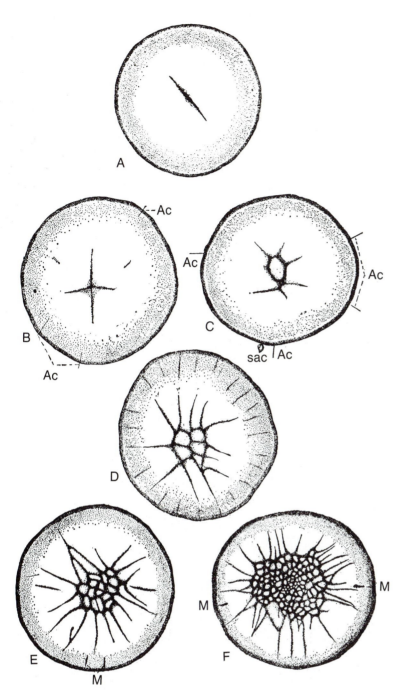

Figure 147 Surface views of the blastoderm of the hen's egg showing cleavage. *a*, surface view of first cleavage furrow in hen's egg (3 hours after fertilization). The dark border represents the inner margin of the periblast. *b*, 4 cells 3¼ hours after fertilization. *c*, 8 cells, 4 hours after fertilization. *d*, 17 cells, 4-5 hours after fertilization. The short radial furrows near the margin are the beginnings of peripheral extensions of primary cleavage furrows. *e*, 34 cells 4¾ hours after fertilization, *f*, 154 cells (in surface view) of which 123 are central cells and 31 marginal; 7 hours after fertilization. The blastoderm at this time averages 3 cells in thickness. ac, accessory cleavage; m, radial furrow; sac, small cell formed by accessory cleavage furrows. (After Patterson.) (From H. L. Wieman, *An Introduction to Vertebrate Embryology*. Copyright 1949 by McGraw-Hill Book Co. Used with permission of McGraw-Hill Book Co.)

11. The Unincubated Chick Blastoderm

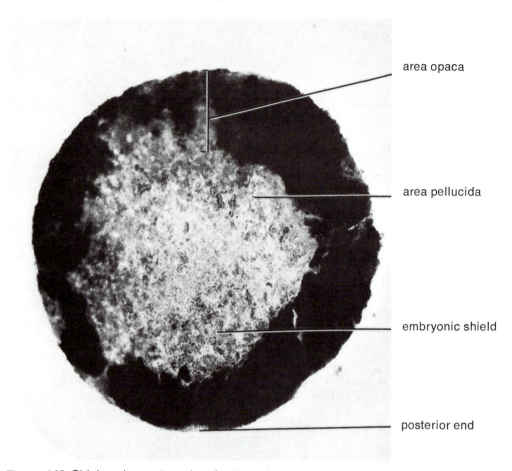

area opaca

area pellucida

embryonic shield

posterior end

Figure 148 Chick embryo, stage 1, unincubated blastoderm, whole mount (mag. 27.5X)

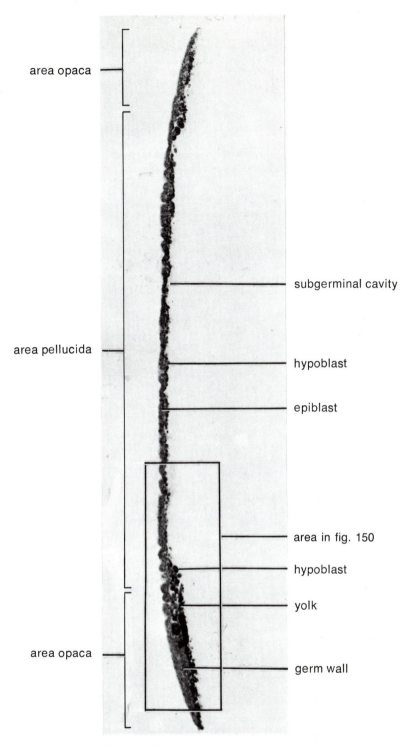

area opaca

area pellucida

area opaca

subgerminal cavity

hypoblast

epiblast

area in fig. 150

hypoblast

yolk

germ wall

Figure 149 Chick embryo, stage 1, unincubated blastoderm, longitudinal section (mag. 50X)

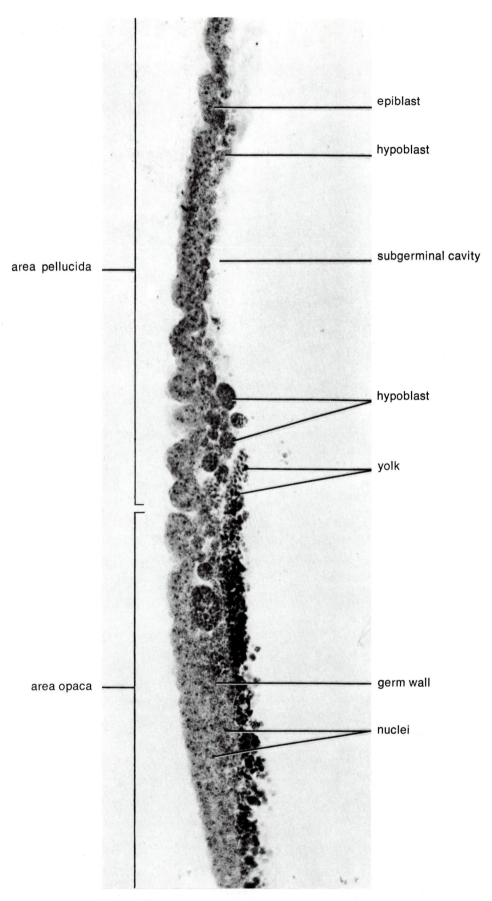

area pellucida

epiblast

hypoblast

subgerminal cavity

hypoblast

yolk

area opaca

germ wall

nuclei

Figure 150 Chick embryo, stage 1, unincubated blastoderm, longitudinal section (mag. 150X)

12. The Notochordal Process Chick Embryo (Stage 5*) (19–22 hours incubation)

*Stages based on Hamburger, V., and H. L. Hamilton, A series of normal stages in the development of the chick embryo, *Journal of Morphology* 88, 1951, pp. 49–92; or see, *Lillie's Development of the Chick* by H. L. Hamilton, Henry Holt and Co., 1952.

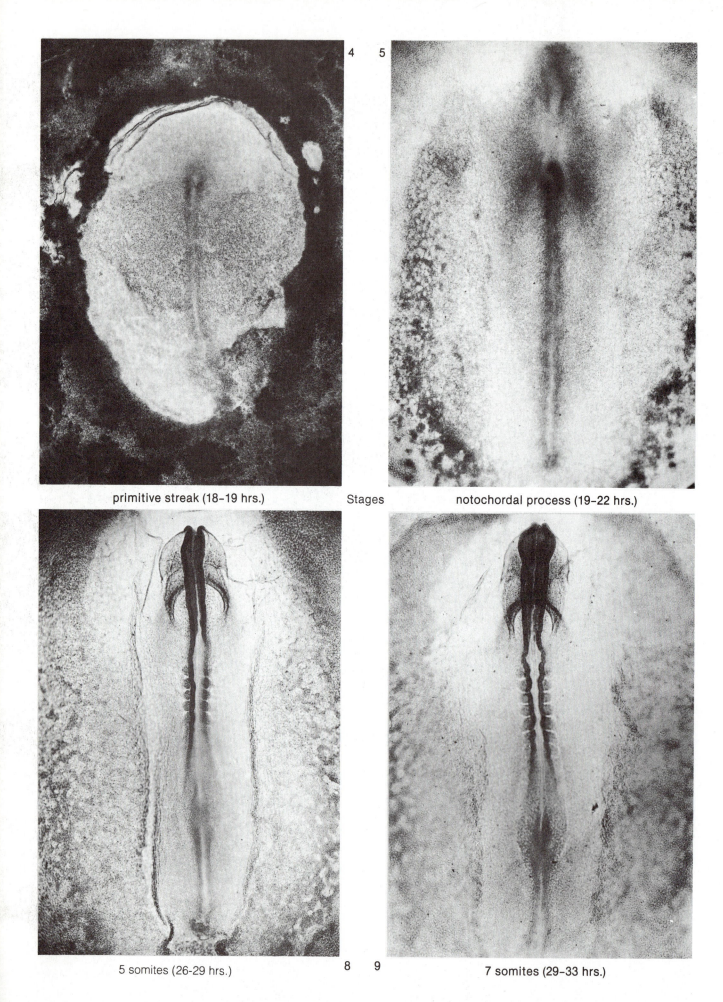

4   5

primitive streak (18–19 hrs.)   Stages   notochordal process (19–22 hrs.)

5 somites (26-29 hrs.)   8   9   7 somites (29–33 hrs.)

Figure 151 Chick embryos, whole mounts (mag. 27.5X)

TABLE 2

NORMAL STAGES
OF CHICK DEVELOPMENT

Because chick embryos develop at different rates under the same environmental conditions, designating developmental stages entirely by incubation times is unsatisfactory. A series of standard or "normal" stages has been established from the structure of embryos. The following brief description of chick embryo stages is based on that of H. L. Hamilton in *Lillie's Development of the Chick*, 3rd ed., 1952, Holt, Rinehart and Winston, New York.[1]

Stage 1. Prestreak (0–5 hours incubation): The embryonic shield may be visible, but no primitive streak has formed. Fig. 148

Stage 2. Initial streak (6–7 hours incubation): The primitive streak appears as a short conical thickening at the border of the area pellucida.

Stage 3. Intermediate streak (12–13 hours): The primitive streak extends to the center of the area pellucida. No primitive groove.

Stage 4. Definitive streak (18–19 hours): The primitive streak is at maximum length, with a primitive groove, primitive pit, and primitive knot present. The area pellucida is pear-shaped and the streak extends two thirds to three fourths of its length. Fig. 151

Stage 5. Head process (19–22 hours): The notochordal process is visible, but no head fold has formed. Fig. 154

Stage 6. Head fold (23–25 hours): The head fold is visible, but no somites have formed.

Stage 7. One somite (23–26 hours): One pair of somites is visible, and neural folds are present.

Stage 8. Four somites (26–29 hours): The neural folds are beginning to close, and blood islands are present. Fig.159, Fig.160

Stage 9. Seven somites (29–33 hours): The optic vesicles are present, and the heart tubes are beginning to fuse.

Stage 10. Ten somites (33–38 hours): Three primary brain vesicles are visible. The optic vesicles are not constricted.

Stage 11. Thirteen somites (40–45 hours): Five neuromeres are visible in the hindbrain. The optic vesicles are constricted at base, and the heart is bent to the right. Fig. 170

Stage 12. Sixteen somites (45–49 hours): the telencephalon is visible; the auditory pits are deep; the heart is S-shaped. The head fold of amnion covers the telencephalon.

Stage 13. Nineteen somites (48–52 hours): The head is turning to the right; the telencephalon is enlarged; the amnion fold covers the head to the hindbrain. Cranial and cervical flexures are present.

Stage 14. Twenty-two somites (50–53 hours): The cranial flexure equals about 90°; visceral arches 1 and 2 and clefts 1 and 2 are distinct; the optic vesicles are invaginated and lens placodes are present.

Stage 15. (50–55 hours): The cranial flexure is more than 90°; visceral arch 3 and cleft 3 are distinct; the optic cup is formed. Fig. 183

Stage 16. (51–56 hours): The wing bud is visible; the tail bud is present; no leg bud is visible.

Stage 17. (52–64 hours): Wing and leg buds are visible; the epiphysis is distinct; nasal pits are forming; the allantois is not visible.

---

[1]Originally published as: V. Hamburger, and H. L. Hamilton, A series of normal stages in the development of the chick embryo. *Journal of morphology*, 88: 49–92, 1951. Used with permission of the Wistar Institute Press.

Stage 18. (3 days): The leg buds are slightly longer than the wing buds; the amnion is nearly or completely closed; the cervical flexure equals about 90°; the maxillary process and 4th cleft are indistinct or absent; the allantois is visible. Fig. 203

Stage 19. (3–3½ days): The maxillary process is distinct and as long as the mandibular process; the allantois is a small pocket but not vesicular; the eye is nonpigmented.

Stage 20. (3–3½ days): The trunk is straight; the allantois is vesicular and about as large as the midbrain; the eye is slightly pigmented. Fig. 224

Stage 21. (3½ days): The limb buds slightly asymmetrical; their axis is directed caudally. The maxillary process is longer than the mandible, extending to the middle of the eye. The 4th arch and cleft are distinct. The allantois extends to the head. Eye pigment is distinct.

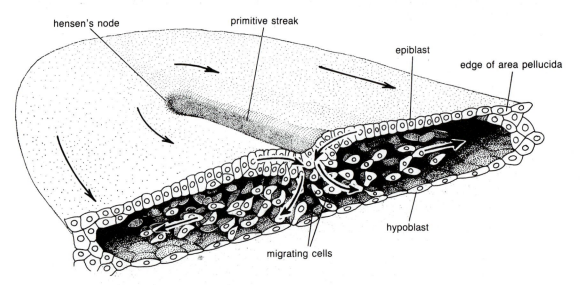

Figure 152 Anterior half of the area pellucida of a chick embryo cut transversely to show the migration of mesodermal and endodermal cells from the primitive streak. (From *An Introduction to Embryology*, 5th edition, by B. I. Balinsky, assisted by B. C. Babian. Copyright © 1981 by CBS College Publishing. Reprinted by permission of W. B. Saunders, a division of CBS College Publishing.)

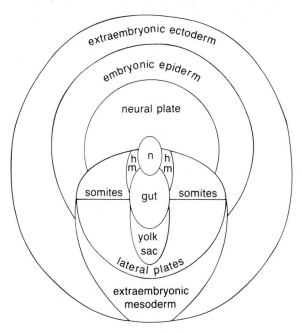

Figure 153 Fate map of the chick embryo at the beginning of gastrulation. (Based on the experiments of G. C. Rosenquist and G. Nicolet.) n, notochord; hm, head mesenchyme. (From *An Introduction to Embryology*, fifth edition by B. I. Balinsky, assisted by B. C. Babian. Copyright © 1981 by CBS College Publishing. Reprinted by permission of W. B. Saunders, a division of CBS College Publishing.)

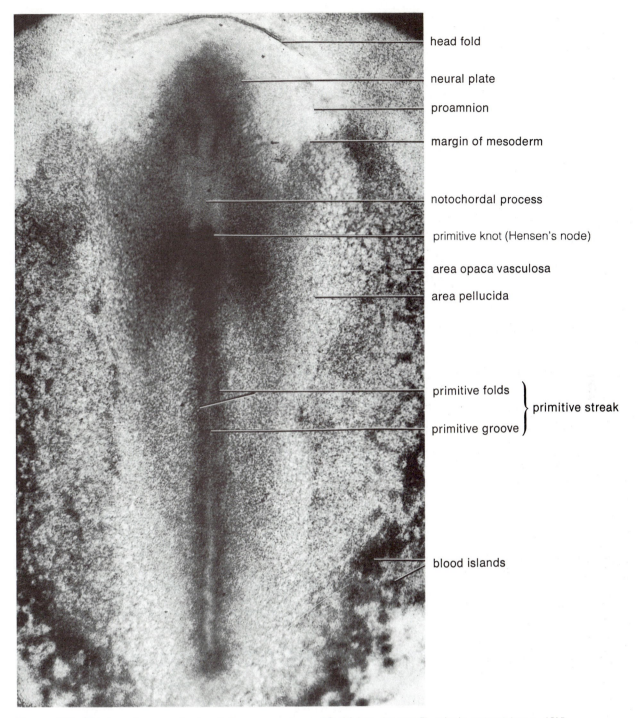

head fold

neural plate

proamnion

margin of mesoderm

notochordal process

primitive knot (Hensen's node)

area opaca vasculosa

area pellucida

primitive folds ⎫
⎬ primitive streak
primitive groove ⎭

blood islands

Figure 154 Chick embryo, notochordal process stage, 19–22 hrs. (stage 5), whole mount (mag. 40X)

Figure 155 Chick embryo, notochordal process stage (stage 5), transverse section through notochordal process (mag. 75X)

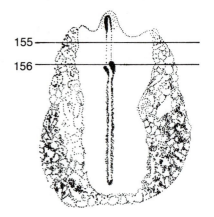

Figure 156 Chick embryo, notochordal process stage (stage 5), transverse section through primitive knot (mag. 75X)

Figure 157 Chick embryo, notochordal process stage (stage 5), transverse section through anterior primitive groove (mag. 75X)

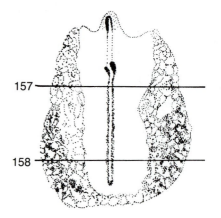

Figure 158 Chick embryo, notochordal process stage (stage 5), transverse section through posterior primitive groove (mag. 75X)

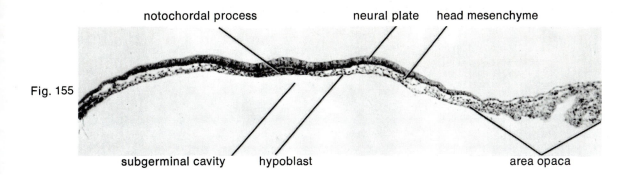

Fig. 155

notochordal process   neural plate   head mesenchyme

subgerminal cavity   hypoblast   area opaca

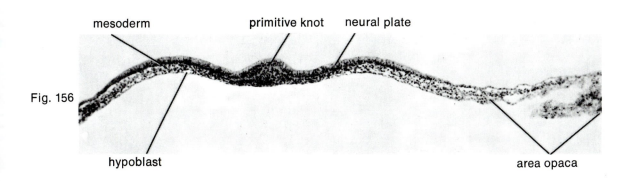

Fig. 156

mesoderm   primitive knot   neural plate

hypoblast   area opaca

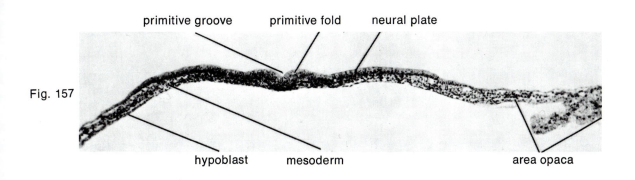

Fig. 157

primitive groove   primitive fold   neural plate

hypoblast   mesoderm   area opaca

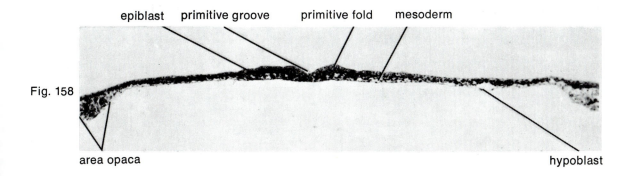

Fig. 158

epiblast   primitive groove   primitive fold   mesoderm

area opaca   hypoblast

13. The 4–5 Somite Chick Embryo (Stage 8) (26–29 hours incubation)

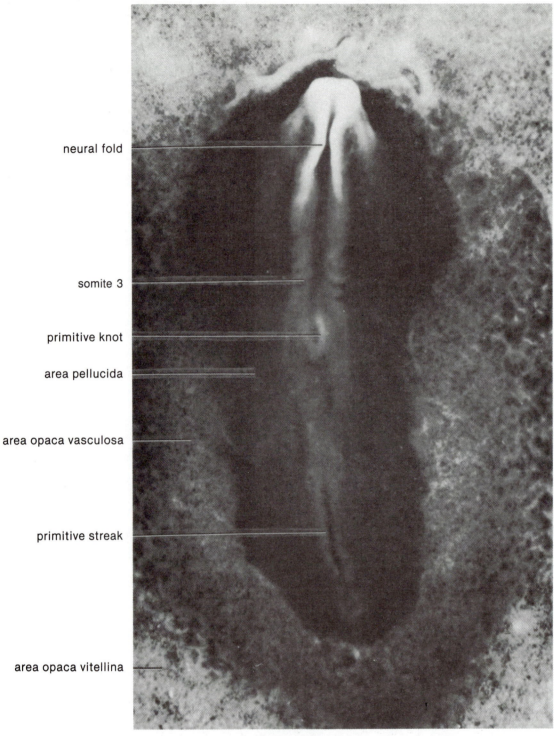

neural fold

somite 3

primitive knot

area pellucida

area opaca vasculosa

primitive streak

area opaca vitellina

Figure 159 4-somite chick embryo (stage 8, opaque whole mount, incident illumination (mag. 40X)

anterior neuropore

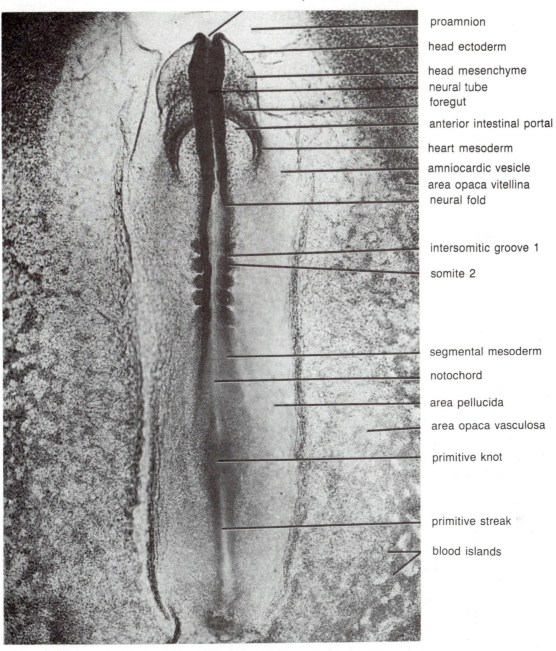

proamnion

head ectoderm

head mesenchyme
neural tube
foregut

anterior intestinal portal

heart mesoderm

amniocardic vesicle
area opaca vitellina
neural fold

intersomitic groove 1

somite 2

segmental mesoderm

notochord

area pellucida

area opaca vasculosa

primitive knot

primitive streak

blood islands

Figure 160 5-somite chick embryo (stage 8), whole mount (mag. 40X)

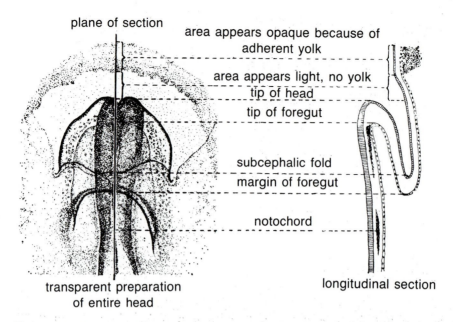

plane of section

area appears opaque because of
adherent yolk

area appears light, no yolk

tip of head

tip of foregut

subcephalic fold

margin of foregut

notochord

transparent preparation
of entire head

longitudinal section

Figure 161 Relation of longitudinal section of the embryonic head of a 24-hour chick embryo to the picture presented by a head of the same age mounted entire as a semitransparent preparation. (From B. M. Patten, *Early Embryology of the Chick*. Copyright 1951 by McGraw-Hill Book Co. Copyright renewed 1957 by B. M. Patten. Used by permission of McGraw-Hill Book Co.)

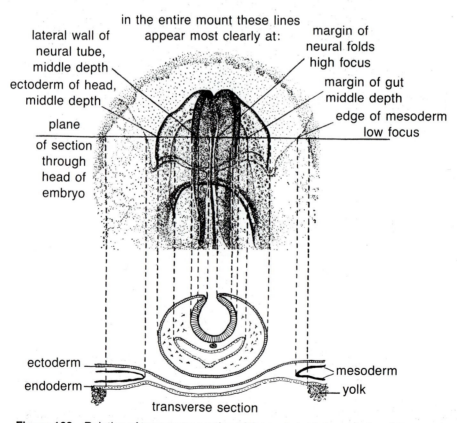

in the entire mount these lines
appear most clearly at:

lateral wall of
neural tube,
middle depth

ectoderm of head,
middle depth

plane

of section

through

head of

embryo

margin of
neural folds
high focus

margin of gut
middle depth

edge of mesoderm
low focus

ectoderm

endoderm

mesoderm

yolk

transverse section

Figure 162 Relation of transverse section of the embryonic head to the picture presented by an entire head of the same age viewed as a semitransparent preparation. The two drawings may be brought into closer relation by looking at them with the top of the page tilted downward. (From B. M. Patten, *Early Embryology of the Chick*. Copyright 1951 McGraw-Hill Book Co. Copyright renewed 1957 by B. M. Patten. Used by permission of McGraw-Hill Book Co.)

126

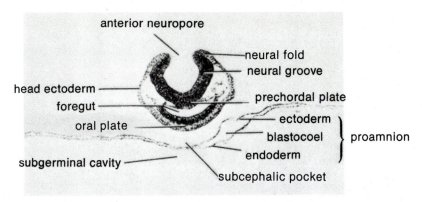

Figure 163 4-somite chick embryo (stage 8), transverse section through pharyngeal membrane (mag. 75X)

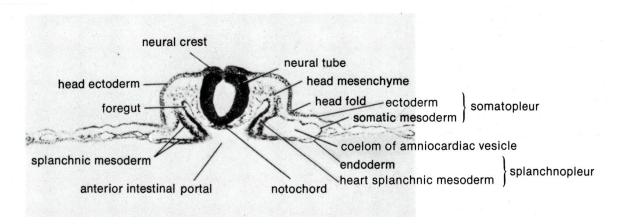

Figure 164 4-somite chick embryo (stage 8), transverse section through anterior intestinal portal (mag. 75X)

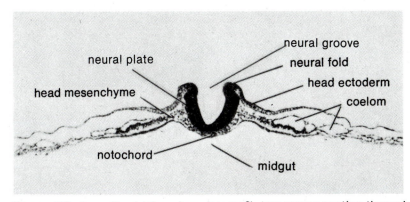

Figure 165 4-somite chick embryo (stage 8), transverse section through neural groove (mag. 75X)

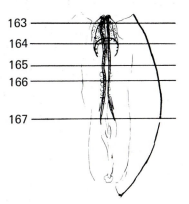

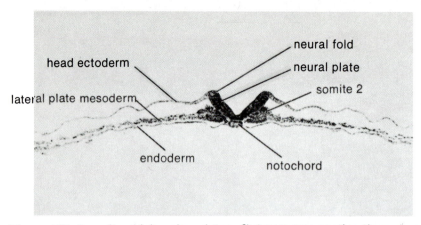

Figure 166 4-somite chick embryo (stage 8), transverse section through somites (mag. 75X)

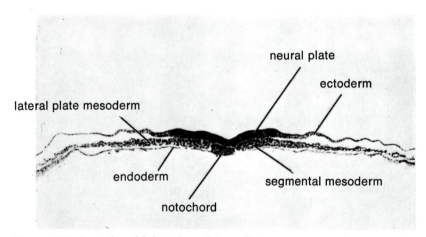

Figure 167 4-somite chick embryo (stage 8), transverse section through neural plate (mag. 75X)

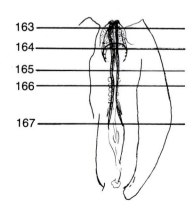

163
164
165
166
167

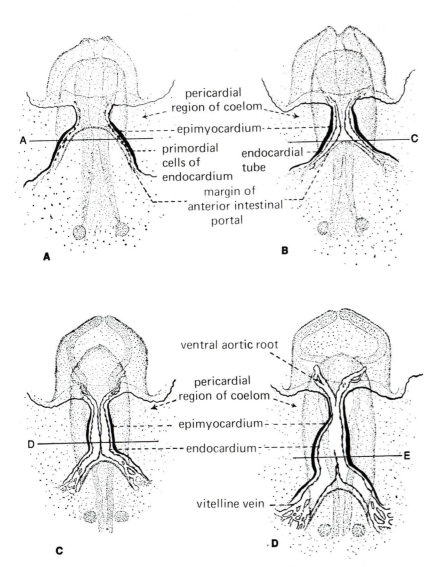

Figure 168 Ventral-view diagrams to show the origin and subsequent fusion of the paired primordia of the heart. The lines A, C, D, and E indicate the planes of the sections diagrammed in fig. 169 *a, c, d,* and *e,* respectively. *a,* chick of 25 hours. *b,* chick of 27 hours. *c,* chick of 28 hours. *d,* chick of 29 hours. (From B. M. Patten, *Early Embryology of the Chick.* Copyright 1951 by McGraw-Hill Book Co. Copyright renewed 1957 by B. M. Patten. Used with permission of McGraw-Hill Book Co.)

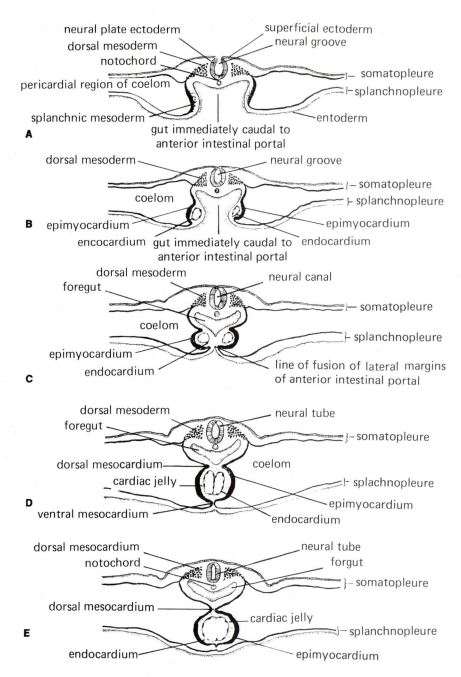

Figure 169 Diagrams of transverse sections through the pericardial region of chicks at various stages to show the formation of the heart. For location of the sections consult fig. 168. *a*, at 25 hours. *b*, at 26 hours. *c*, at 27 hours. *d*, at 28 hours. *e*, at 29 hours. (From B. M. Patten, *Early Embryology of the Chick*. Copyright 1951 by McGraw-Hill Book Co. Copyright renewed 1957 by B. M. Patten. Used with permission of McGraw-Hill Book Co.)

14. The 13-Somite Chick Embryo (Stage 11) (40–45 hours incubation)

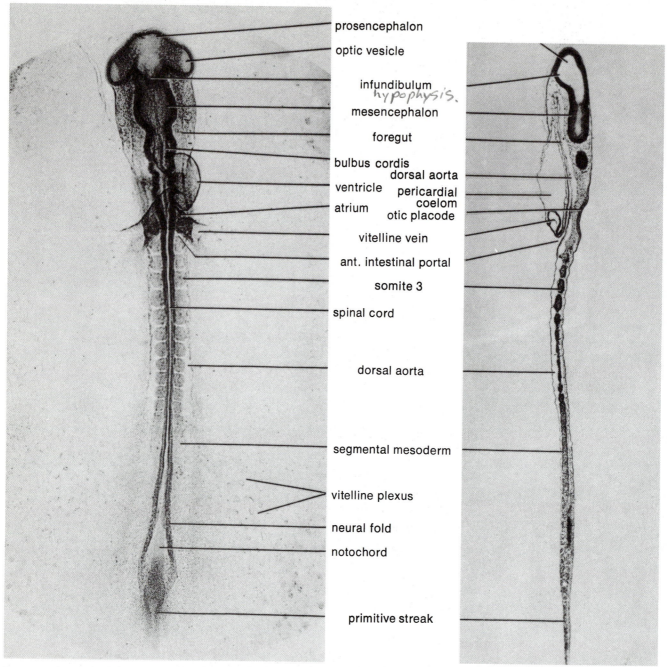

prosencephalon

optic vesicle

infundibulum
hypophysis.

mesencephalon

foregut

bulbus cordis
dorsal aorta

ventricle
pericardial
coelom

atrium
otic placode

vitelline vein

ant. intestinal portal

somite 3

spinal cord

dorsal aorta

segmental mesoderm

vitelline plexus

neural fold

notochord

primitive streak

Figure 170 12-somite chick embryo (stage 11), whole mount (mag. 30X)

Figure 171 12-somite chick embryo (stage 11), sagittal section (mag. 30X)

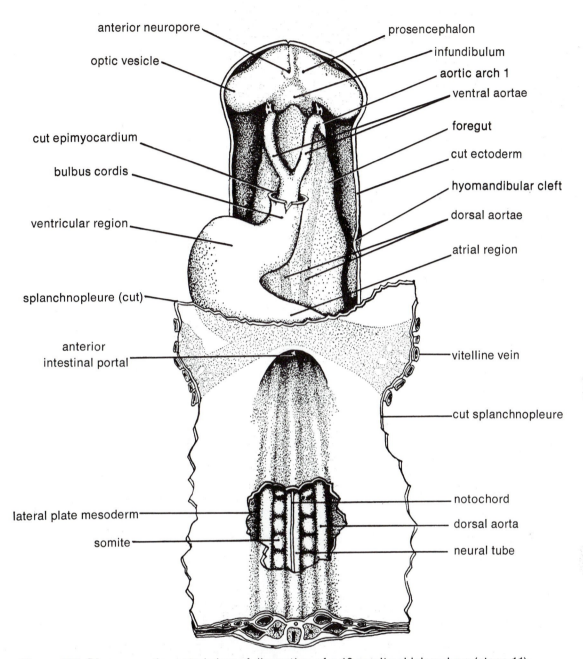

Figure 172 Diagrammatic ventral view of dissection of a 13-somite chick embryo (stage 11). (Modified from Prentiss.) The splanchnopleure of the yolk sac cephalic to the anterior intestinal portal, the ectoderm of the ventral surface of the head, and the mesoderm of the pericardial region, have been removed to show the underlying structures. (From *Early Embryology of the Chick,* 4th ed., by Bradley M. Patten. Copyright 1951 by McGraw-Hill Book Co. Copyright renewed 1957 by B. M. Patten. Used with permission of the McGraw-Hill Book Co.)

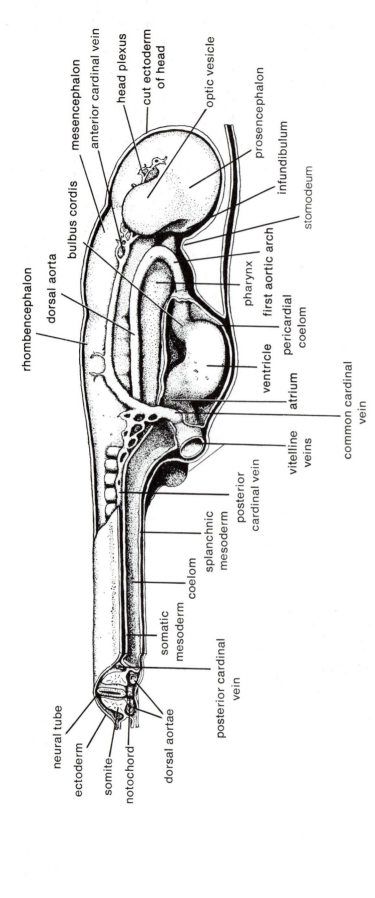

Figure 173 Diagrammatic lateral view of dissection of a 12-somite chick (stage 11). The lateral body wall on the right side has been removed to show the internal structures. Note especially the relations of the pericardial region to that part of the coelom which lies farther caudally, and the small anastomosing channels of the developing posterior cardinal vein from which a single main vessel is later derived. (From *Early Embryology of the Chick*, 4th ed., by Bradley M. Patten. Copyright 1951 by McGraw-Hill Book Co. Copyright renewed 1957 by Bradley M. Patten. Used with permission of the McGraw-Hill Book Co.)

Labels: rhombencephalon, dorsal aorta, bulbus cordis, mesencephalon, anterior cardinal vein, head plexus, cut ectoderm of head, optic vesicle, prosencephalon, infundibulum, stomodeum, first aortic arch, pharynx, pericardial coelom, ventricle, atrium, common cardinal vein, vitelline veins, posterior cardinal vein, splanchnic mesoderm, coelom, somatic mesoderm, posterior cardinal vein, dorsal aortae, notochord, somite, ectoderm, neural tube

134

15. The 2-Day Chick Embryo (Stage 15) (50–55 hours incubation)

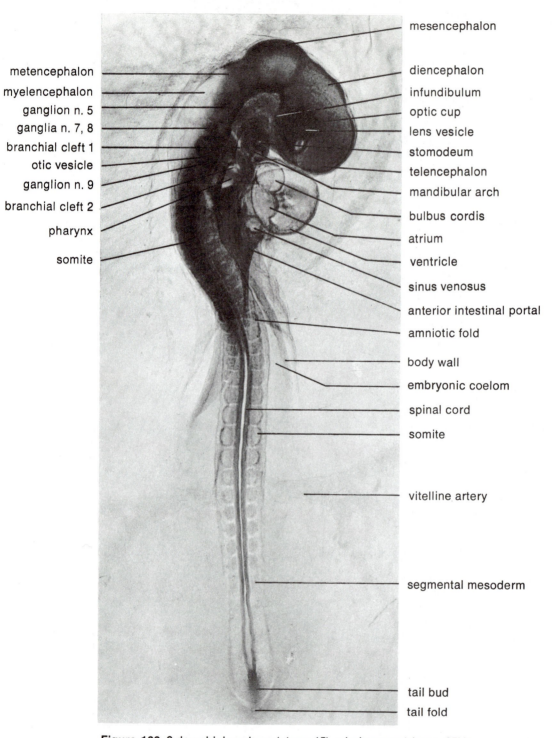

mesencephalon

metencephalon

myelencephalon

ganglion n. 5

ganglia n. 7, 8

branchial cleft 1

otic vesicle

ganglion n. 9

branchial cleft 2

pharynx

somite

diencephalon

infundibulum

optic cup

lens vesicle

stomodeum

telencephalon

mandibular arch

bulbus cordis

atrium

ventricle

sinus venosus

anterior intestinal portal

amniotic fold

body wall

embryonic coelom

spinal cord

somite

vitelline artery

segmental mesoderm

tail bud

tail fold

Figure 183 2-day chick embryo (stage 15), whole mount (mag. 25X)

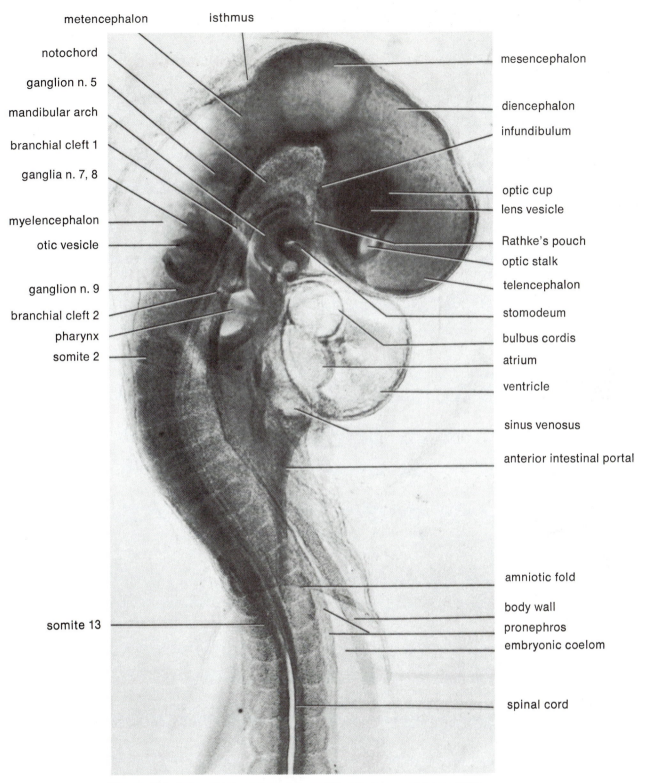

metencephalon

isthmus

notochord

ganglion n. 5

mandibular arch

branchial cleft 1

ganglia n. 7, 8

myelencephalon

otic vesicle

ganglion n. 9

branchial cleft 2

pharynx

somite 2

somite 13

mesencephalon

diencephalon

infundibulum

optic cup

lens vesicle

Rathke's pouch

optic stalk

telencephalon

stomodeum

bulbus cordis

atrium

ventricle

sinus venosus

anterior intestinal portal

amniotic fold

body wall

pronephros

embryonic coelom

spinal cord

Figure 184 2-day chick embryo (stage 15), whole mount (mag. 50X)

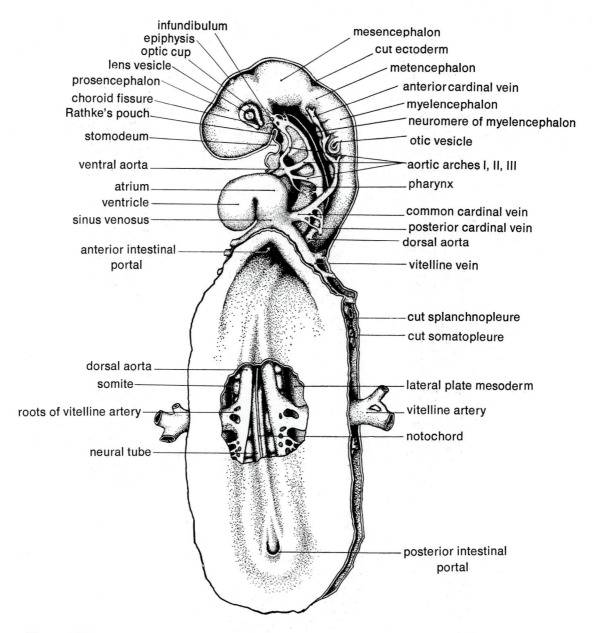

Figure 185

Diagram of dissection of chick of about 50 hours. (Modified from Prentiss.) The splanchnopleure of the yolk sac cephalic to the anterior intestinal portal, the ectoderm of the left side of the head, and the mesoderm in the pericardial region have been dissected away. A window has been cut in the splanchnopleure of the dorsal wall of the mid gut to show the origin of the vitelline artery. (From *Early Embryology of the Chick*, 4th ed., by Bradley M. Patten. Copyright 1951 by McGraw-Hill Book Co. Copyright renewed 1957 by B. M. Patten. Used with permission of the McGraw-Hill Book Co.)

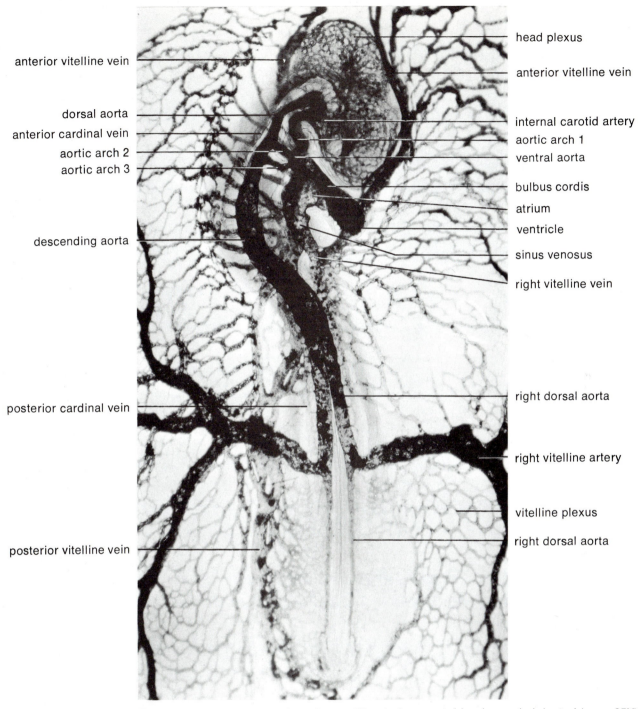

anterior vitelline vein

head plexus

anterior vitelline vein

dorsal aorta
anterior cardinal vein
aortic arch 2
aortic arch 3

internal carotid artery
aortic arch 1
ventral aorta

descending aorta

bulbus cordis
atrium
ventricle
sinus venosus
right vitelline vein

posterior cardinal vein

right dorsal aorta

right vitelline artery

vitelline plexus
right dorsal aorta

posterior vitelline vein

Figure 186 2-day chick embryo (stage 15), whole mount, blood vessels injected (mag. 25X)

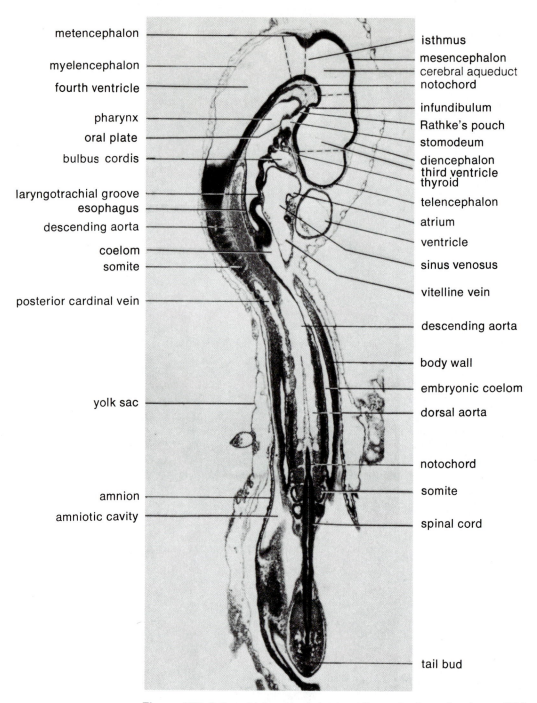

metencephalon

myelencephalon

fourth ventricle

pharynx

oral plate

bulbus cordis

laryngotrachial groove
esophagus

descending aorta

coelom

somite

posterior cardinal vein

yolk sac

amnion

amniotic cavity

isthmus

mesencephalon
cerebral aqueduct
notochord

infundibulum

Rathke's pouch

stomodeum

diencephalon
third ventricle
thyroid

telencephalon

atrium

ventricle

sinus venosus

vitelline vein

descending aorta

body wall

embryonic coelom

dorsal aorta

notochord

somite

spinal cord

tail bud

Figure 187 2-day chick embryo (stage 15), sagittal section (mag. 25X)

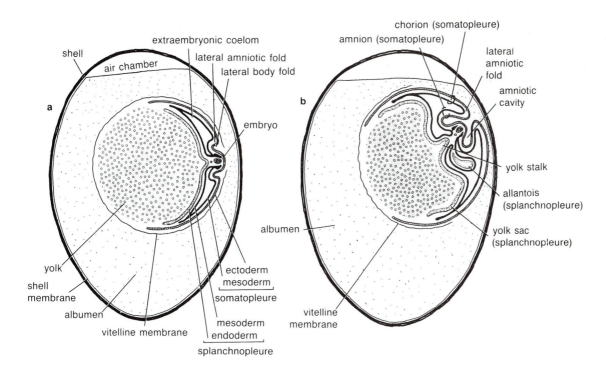

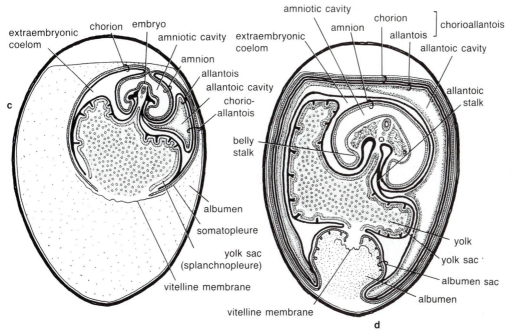

Figure 188 Schematic diagrams to show the extraembryonic membranes of the chick. (After Duval.) The diagrams represent longitudinal sections through the entire egg. The body of the embryo, being oriented approximately at right angles to the long axis of the egg, is cut transversely. *a*, embryo of about 2 days incubation. *b*, embryo of about 3 days incubation. *c*, embryo of about 5 days incubation. *d*, embryo of about 14 days incubation. (From B. M. Patten, *Early Embryology of the Chick.* Copyright 1951 by McGraw-Hill Book Co. Copyright renewed 1957 by B. M. Patten. Used with permission of McGraw-Hill Book Co.)

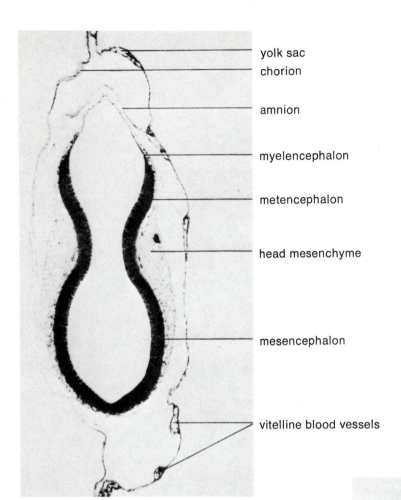

yolk sac

chorion

amnion

myelencephalon

metencephalon

head mesenchyme

mesencephalon

vitelline blood vessels

Figure 189 2-day chick embryo (stage 15), transverse section through mesencephalon (mag. 60X)

Figure 190 2-day chick embryo (stage 15), transverse section through ganglion of cranial nerve 5 (mag. 60X)

189
190

myelencephalon

ganglion n. 5

anterior cardinal vein

head mesenchyme

diencephalon

amniotic cavity

amnion
extraembryonic coelom

chorion

yolk sac

146

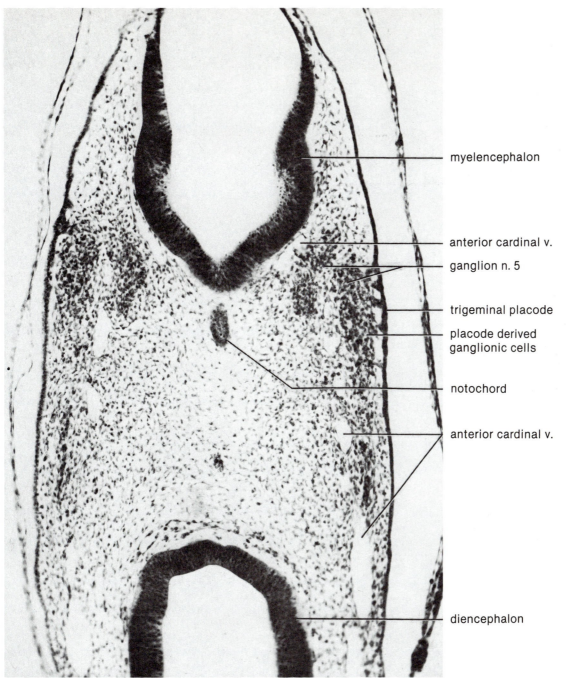

myelencephalon

anterior cardinal v.

ganglion n. 5

trigeminal placode

placode derived
ganglionic cells

notochord

anterior cardinal v.

diencephalon

Figure 191 2-day chick embryo (stage 15), transverse section through trigeminal placode (mag. 135X)

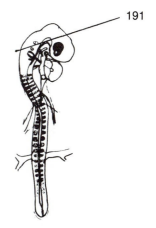

191

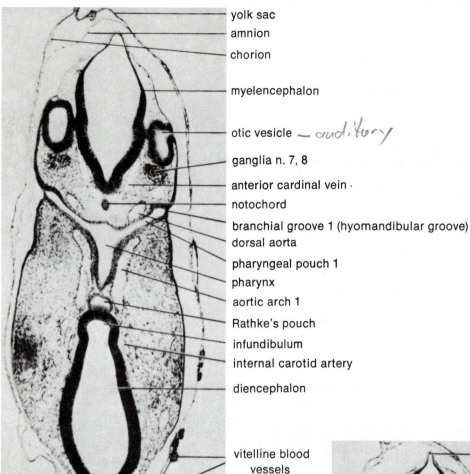

yolk sac
amnion
chorion

myelencephalon

otic vesicle _ auditory

ganglia n. 7, 8

anterior cardinal vein ·

notochord

branchial groove 1 (hyomandibular groove)

dorsal aorta

pharyngeal pouch 1

pharynx

aortic arch 1

Rathke's pouch

infundibulum

internal carotid artery

diencephalon

vitelline blood
vessels

Figure 192 2-day chick embryo (stage 15),
transverse section through otic
vesicle (mag. 60X)

Figure 193 2-day chick embryo (stage 15),
transverse section through optic
cups (mag. 60X)

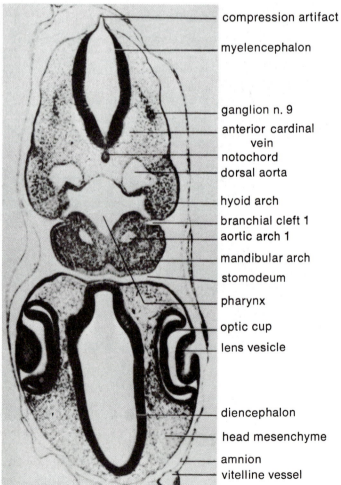

compression artifact

myelencephalon

ganglion n. 9

anterior cardinal
vein

notochord

dorsal aorta

hyoid arch

branchial cleft 1

aortic arch 1

mandibular arch

stomodeum

pharynx

optic cup

lens vesicle

diencephalon

head mesenchyme

amnion

vitelline vessel

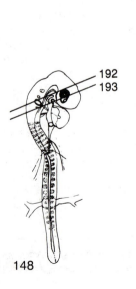

192
193

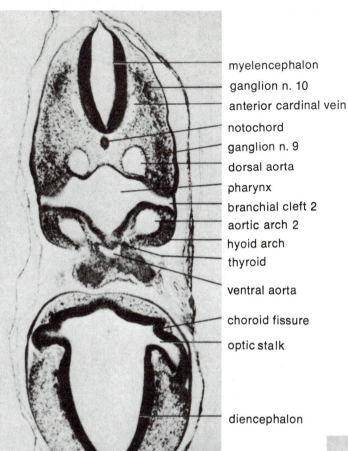

myelencephalon
ganglion n. 10
anterior cardinal vein
notochord
ganglion n. 9
dorsal aorta
pharynx
branchial cleft 2
aortic arch 2
hyoid arch
thyroid

ventral aorta

choroid fissure

optic stalk

diencephalon

Figure 194 2-day chick embryo (stage 15), transverse section through thyroid (mag. 60X)

Figure 195
2-day chick embryo (stage 15), transverse section through olfactory placodes (mag. 60X)

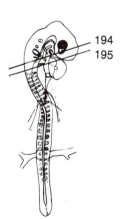

194
195

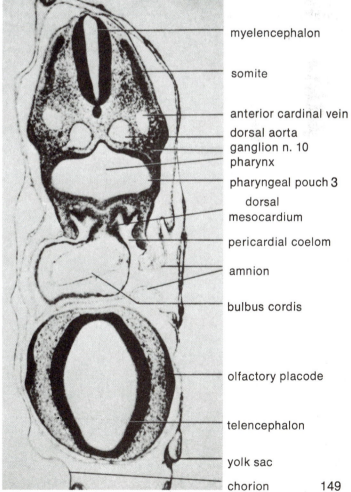

myelencephalon

somite

anterior cardinal vein
dorsal aorta
ganglion n. 10
pharynx

pharyngeal pouch 3

dorsal mesocardium

pericardial coelom

amnion

bulbus cordis

olfactory placode

telencephalon

yolk sac

chorion

149

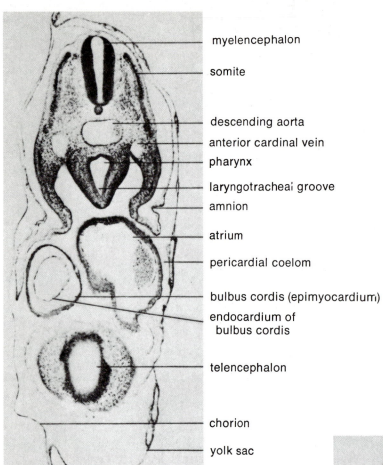

myelencephalon

somite

descending aorta

anterior cardinal vein

pharynx

laryngotracheal groove

amnion

atrium

pericardial coelom

bulbus cordis (epimyocardium)

endocardium of
 bulbus cordis

telencephalon

chorion

yolk sac

Figure 196 2-day chick embryo (stage 15), transverse
section through atrium (mag. 60X)

Figure 197
2-day chick embryo (stage 15), transverse
section through sinus venosus
(mag. 60X)

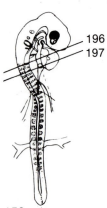

196
197

150

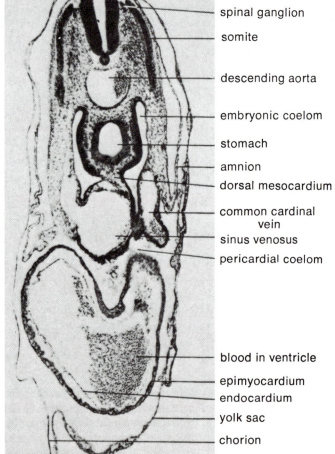

spinal cord

spinal ganglion

somite

descending aorta

embryonic coelom

stomach

amnion

dorsal mesocardium

common cardinal
vein

sinus venosus

pericardial coelom

blood in ventricle

epimyocardium

endocardium

yolk sac

chorion

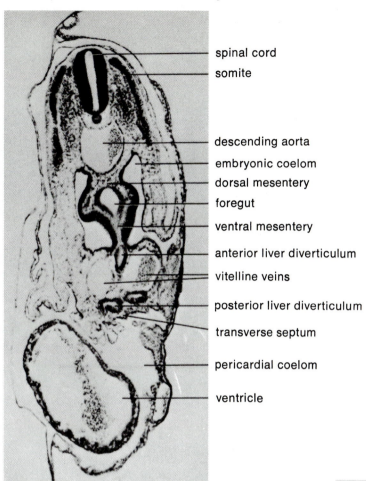

spinal cord
somite

descending aorta
embryonic coelom
dorsal mesentery
foregut
ventral mesentery
anterior liver diverticulum
vitelline veins
posterior liver diverticulum
transverse septum
pericardial coelom
ventricle

Figure 198 2-day chick embryo (stage 15), transverse section through liver diverticula (mag. 60X)

Figure 199
2-day chick embryo (stage 15), transverse section through anterior intestinal portal (mag. 60X)

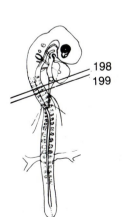

198
199

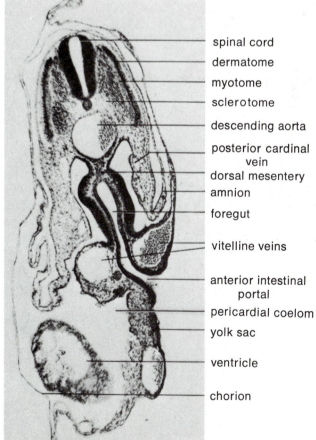

spinal cord
dermatome
myotome
sclerotome
descending aorta
posterior cardinal vein
dorsal mesentery
amnion
foregut
vitelline veins
anterior intestinal portal
pericardial coelom
yolk sac
ventricle
chorion

151

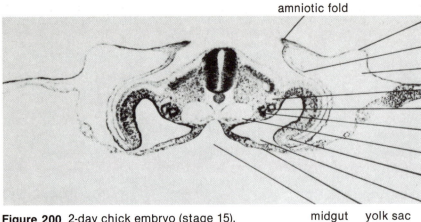

amniotic fold
chorion
amnion
extraembryonic coelom
posterior cardinal vein
pronephric duct
pronephric tubule
dorsal aorta
lateral body fold
embryonic coelom
midgut   yolk sac

Figure 200 2-day chick embryo (stage 15), transverse section through pronephros (mag. 60X)

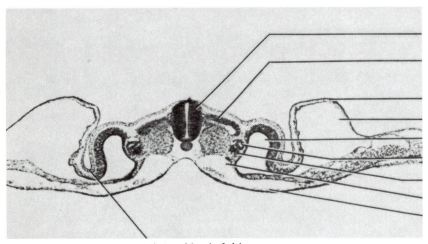

spinal cord
somite
amniotic fold
extraembryonic coelom
embryonic coelom
mesonephric duct
nephrotome
dorsal aorta
vitelline artery
lateral body fold

Figure 201
2-day chick embryo (stage 15)
transverse section through vitelline artery
(mag. 60X)

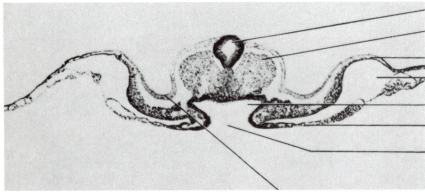

spinal cord
segmental mesoderm
amniotic fold
coelom
hindgut
yolk sac
posterior intestinal portal
tail fold

Figure 202 2-day chick embryo (stage 15, transverse section through posterior intestinal portal (mag. 60X)

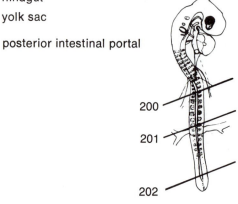

200

201

202

16. The 3-Day Chick Embryo (Stage 18)

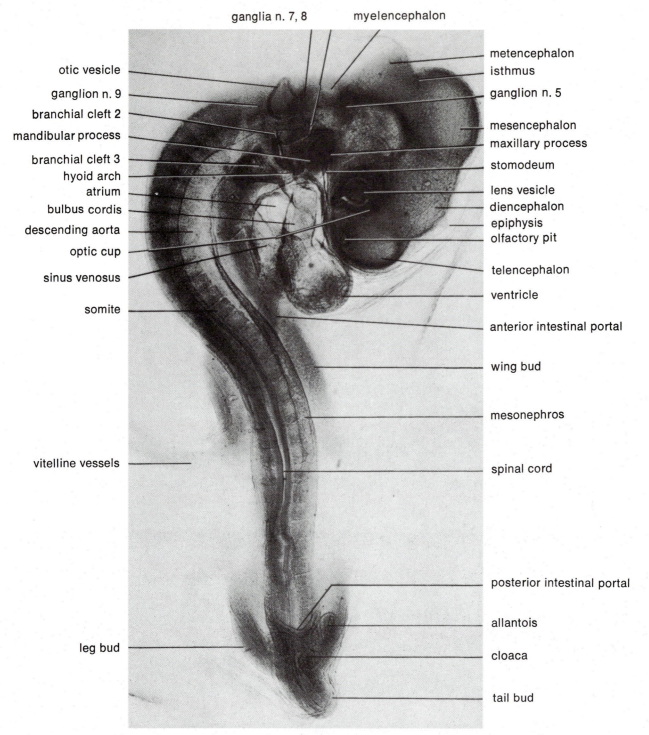

branchial cleft 1

ganglia n. 7, 8   myelencephalon

metencephalon
isthmus

otic vesicle
ganglion n. 9
branchial cleft 2
mandibular process
branchial cleft 3
hyoid arch
atrium
bulbus cordis
descending aorta
optic cup
sinus venosus

somite

vitelline vessels

leg bud

ganglion n. 5

mesencephalon
maxillary process
stomodeum

lens vesicle
diencephalon
epiphysis
olfactory pit

telencephalon

ventricle

anterior intestinal portal

wing bud

mesonephros

spinal cord

posterior intestinal portal

allantois

cloaca

tail bud

Figure 203 3-day chick embryo (stage 18), whole mount
(mag. 25X; transmitted illumination)

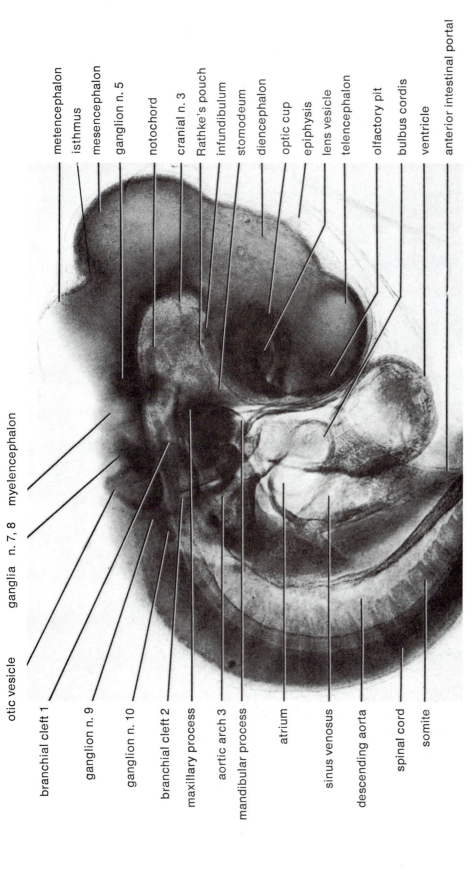

otic vesicle

ganglia n. 7, 8   myelencephalon

branchial cleft 1

ganglion n. 9

ganglion n. 10

branchial cleft 2

maxillary process

aortic arch 3

mandibular process

atrium

sinus venosus

descending aorta

spinal cord

somite

metencephalon

isthmus

mesencephalon

ganglion n. 5

notochord

cranial n. 3

Rathke's pouch

infundibulum

stomodeum

diencephalon

optic cup

epiphysis

lens vesicle

telencephalon

olfactory pit

bulbus cordis

ventricle

anterior intestinal portal

Figure 204
3-day chick embryo (stage 18), whole mount (mag. 50 X)

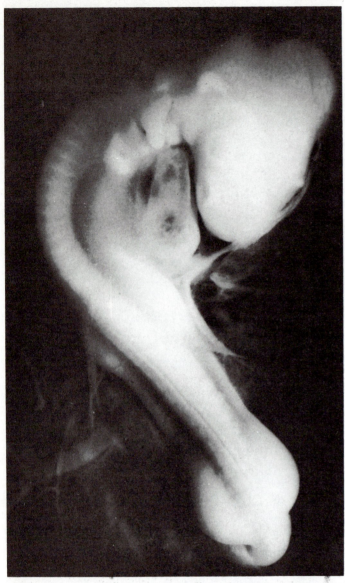

Figure 205 3-day chick embryo (stage 19) opaque mount, unstained, incident illumination

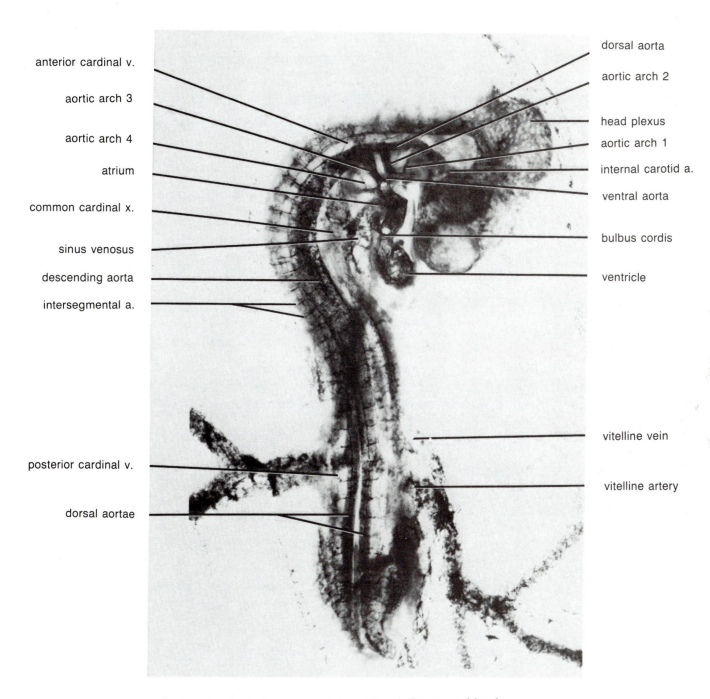

anterior cardinal v.

aortic arch 3

aortic arch 4

atrium

common cardinal x.

sinus venosus

descending aorta

intersegmental a.

posterior cardinal v.

dorsal aortae

dorsal aorta

aortic arch 2

head plexus

aortic arch 1

internal carotid a.

ventral aorta

bulbus cordis

ventricle

vitelline vein

vitelline artery

Figure 206 3-day chick embryo (stage 18), whole mount, blood
vessels injected (mag. 25X; transmitted illumination)

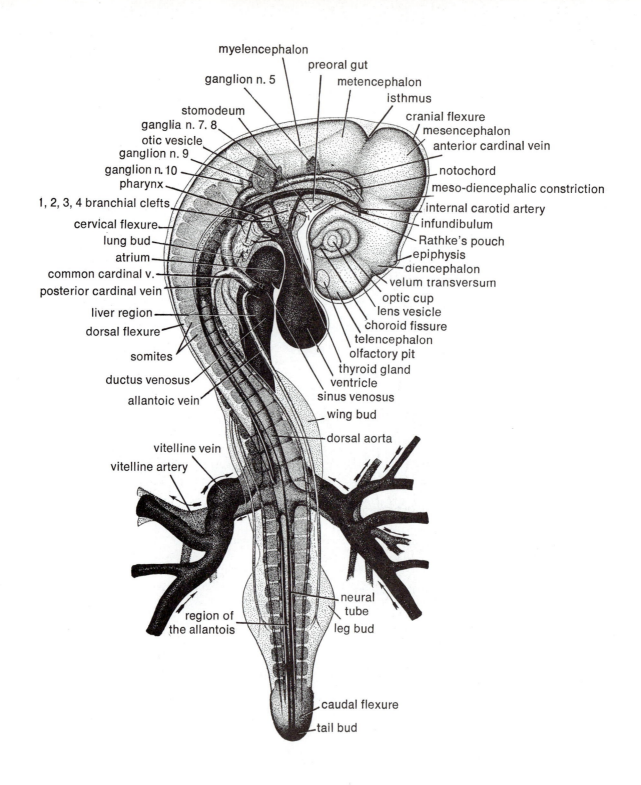

Figure 207
The 3-day chick embryo with 35 somites (stage 18), dorsal view. (From *Fundamentals of Comparative Embryology of the Vertebrates* by Alfred F. Huettner. Copyright 1941 by Macmillan Publishing Company, New York. Used with permission of Macmillan Publishing Company.)

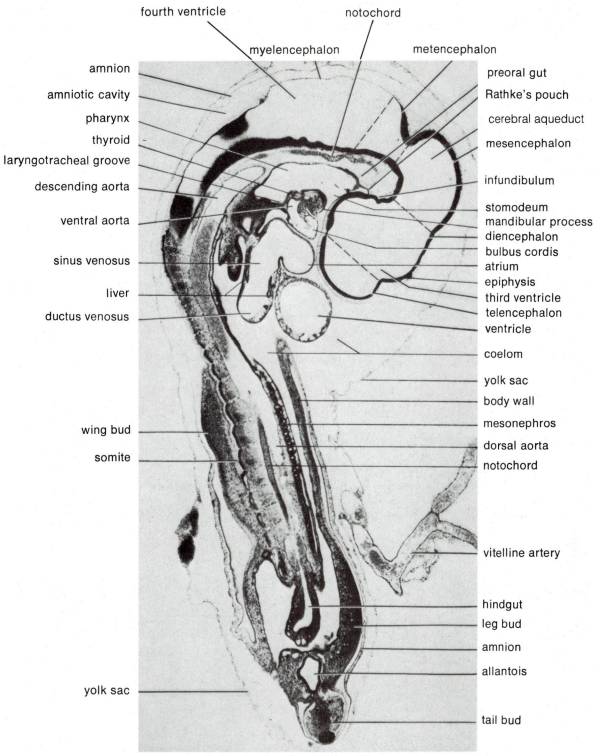

fourth ventricle

notochord

myelencephalon

metencephalon

amnion

amniotic cavity

pharynx

thyroid

laryngotracheal groove

descending aorta

ventral aorta

sinus venosus

liver

ductus venosus

wing bud

somite

yolk sac

preoral gut

Rathke's pouch

cerebral aqueduct

mesencephalon

infundibulum

stomodeum

mandibular process

diencephalon

bulbus cordis

atrium

epiphysis

third ventricle

telencephalon

ventricle

coelom

yolk sac

body wall

mesonephros

dorsal aorta

notochord

vitelline artery

hindgut

leg bud

amnion

allantois

tail bud

Figure 208 3-day chick embryo (stage 18), sagittal section (mag. 25X)

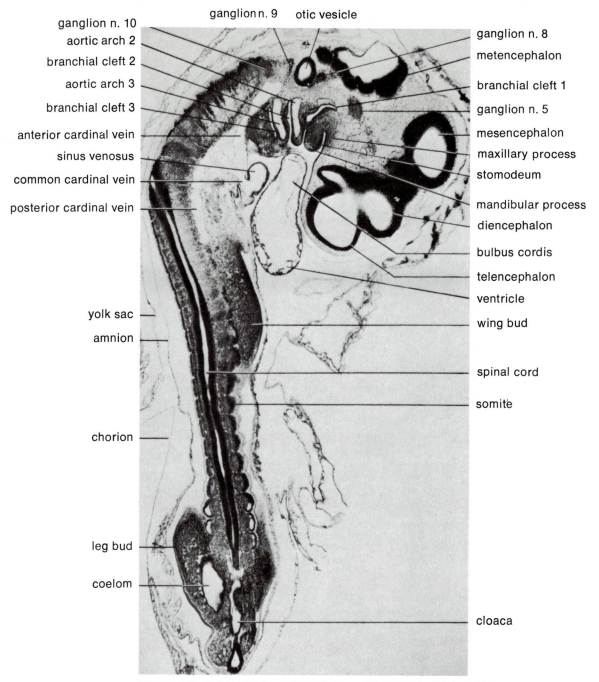

ganglion n. 10
aortic arch 2
branchial cleft 2
aortic arch 3
branchial cleft 3
anterior cardinal vein
sinus venosus
common cardinal vein
posterior cardinal vein
yolk sac
amnion
chorion
leg bud
coelom

ganglion n. 9   otic vesicle

ganglion n. 8
metencephalon
branchial cleft 1
ganglion n. 5
mesencephalon
maxillary process
stomodeum
mandibular process
diencephalon
bulbus cordis
telencephalon
ventricle
wing bud
spinal cord
somite
cloaca

Figure 209 3-day chick embryo (stage 18), parasagittal section, right side (mag. 25X)

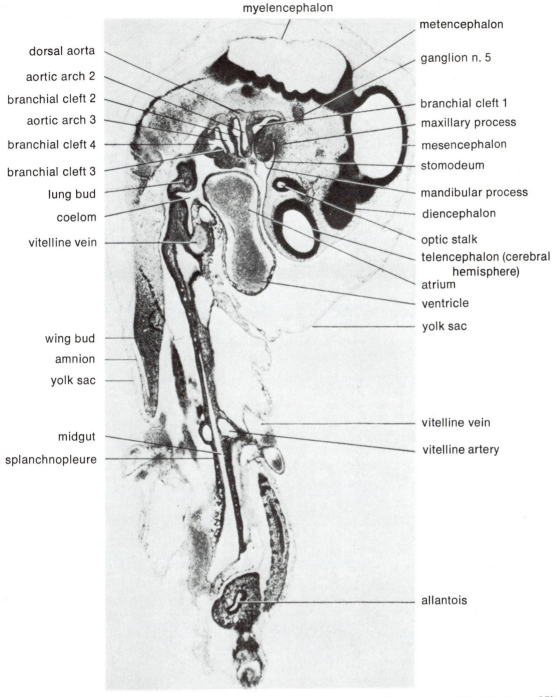

myelencephalon

metencephalon

ganglion n. 5

dorsal aorta

aortic arch 2

branchial cleft 2

aortic arch 3

branchial cleft 4

branchial cleft 3

lung bud

coelom

vitelline vein

branchial cleft 1

maxillary process

mesencephalon

stomodeum

mandibular process

diencephalon

optic stalk

telencephalon (cerebral hemisphere)

atrium

ventricle

yolk sac

wing bud

amnion

yolk sac

vitelline vein

vitelline artery

midgut

splanchnopleure

allantois

Figure 210 3-day chick embryo (stage 18), parasagittal section, left side (mag. 25X)

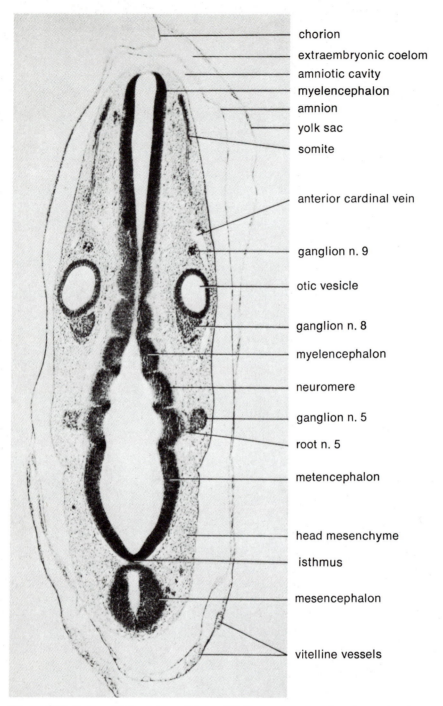

Figure 211 3-day chick embryo (stage 18), transverse section through otic vesicles (mag. 50X)

The labels for the figure, from top to bottom, are:

chorion
extraembryonic coelom
amniotic cavity
myelencephalon
amnion
yolk sac
somite

anterior cardinal vein

ganglion n. 9

otic vesicle

ganglion n. 8

myelencephalon

neuromere

ganglion n. 5

root n. 5

metencephalon

head mesenchyme

isthmus

mesencephalon

vitelline vessels

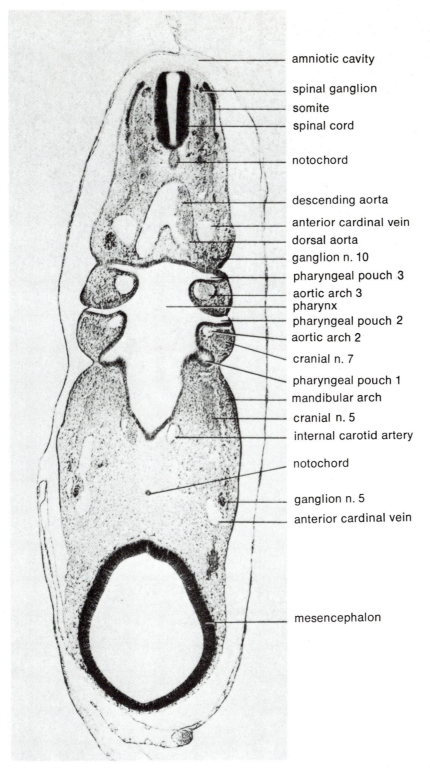

amniotic cavity

spinal ganglion

somite

spinal cord

notochord

descending aorta

anterior cardinal vein

dorsal aorta

ganglion n. 10

pharyngeal pouch 3

aortic arch 3

pharynx

pharyngeal pouch 2

aortic arch 2

cranial n. 7

pharyngeal pouch 1

mandibular arch

cranial n. 5

internal carotid artery

notochord

ganglion n. 5

anterior cardinal vein

mesencephalon

Figure 212 3-day chick embryo (stage 18), transverse section through pharynx (mag. 50X)

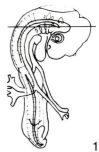

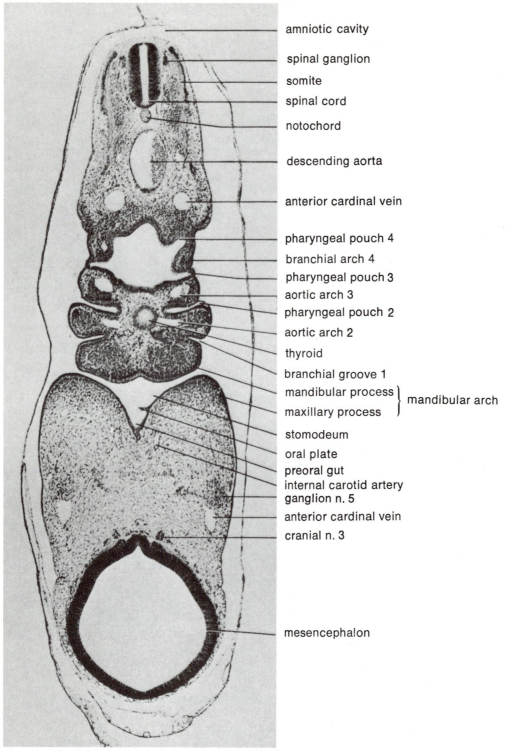

amniotic cavity

spinal ganglion

somite

spinal cord

notochord

descending aorta

anterior cardinal vein

pharyngeal pouch 4

branchial arch 4

pharyngeal pouch 3

aortic arch 3

pharyngeal pouch 2

aortic arch 2

thyroid

branchial groove 1

mandibular process ⎱
⎰ mandibular arch

maxillary process

stomodeum

oral plate

preoral gut

internal carotid artery

ganglion n. 5

anterior cardinal vein

cranial n. 3

mesencephalon

Figure 213 3-day chick embryo (stage 18), transverse section through thyroid (mag. 50X)

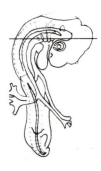

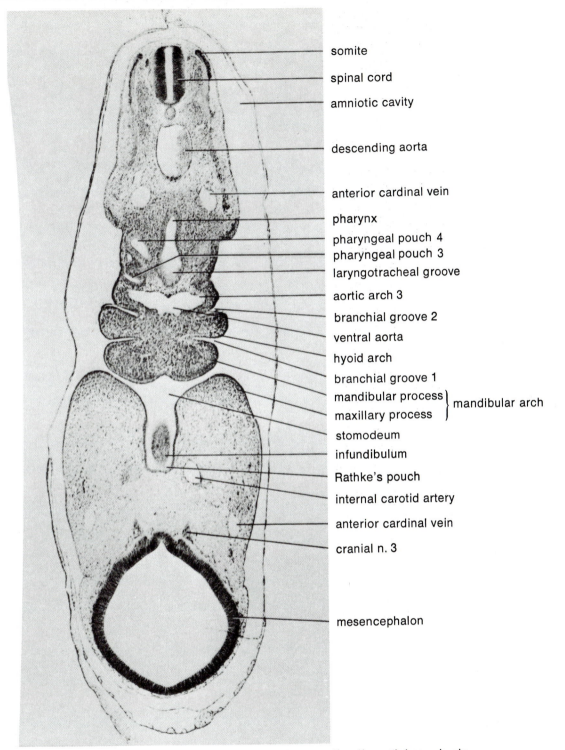

somite

spinal cord

amniotic cavity

descending aorta

anterior cardinal vein

pharynx

pharyngeal pouch 4

pharyngeal pouch 3

laryngotracheal groove

aortic arch 3

branchial groove 2

ventral aorta

hyoid arch

branchial groove 1

mandibular process ⎫
                    ⎬ mandibular arch
maxillary process ⎭

stomodeum

infundibulum

Rathke's pouch

internal carotid artery

anterior cardinal vein

cranial n. 3

mesencephalon

Figure 214 3-day chick embryo (stage 18), transverse section through hypophysis
(mag. 50X)

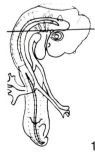

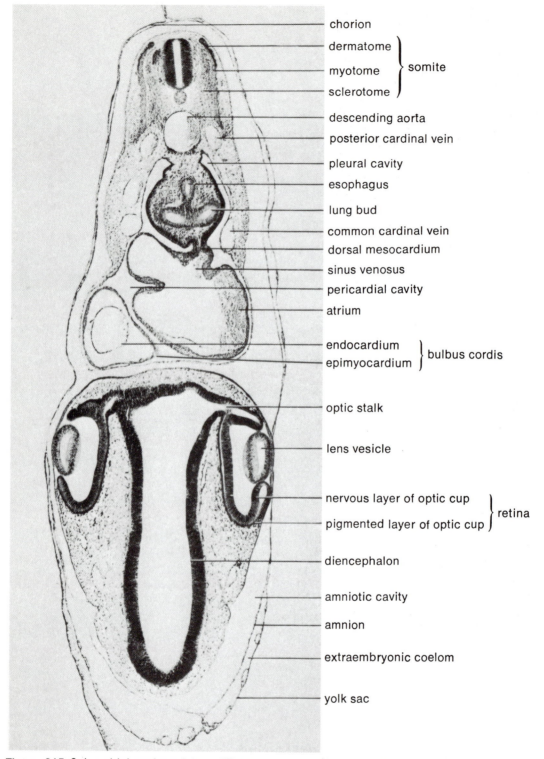

chorion

dermatome ⎫
myotome ⎬ somite
sclerotome ⎭

descending aorta

posterior cardinal vein

pleural cavity

esophagus

lung bud

common cardinal vein

dorsal mesocardium

sinus venosus

pericardial cavity

atrium

endocardium ⎫
epimyocardium ⎬ bulbus cordis

optic stalk

lens vesicle

nervous layer of optic cup ⎫
pigmented layer of optic cup ⎬ retina

diencephalon

amniotic cavity

amnion

extraembryonic coelom

yolk sac

Figure 215 3-day chick embryo (stage 18), transverse section through optic cups
(mag. 50X)

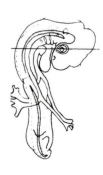

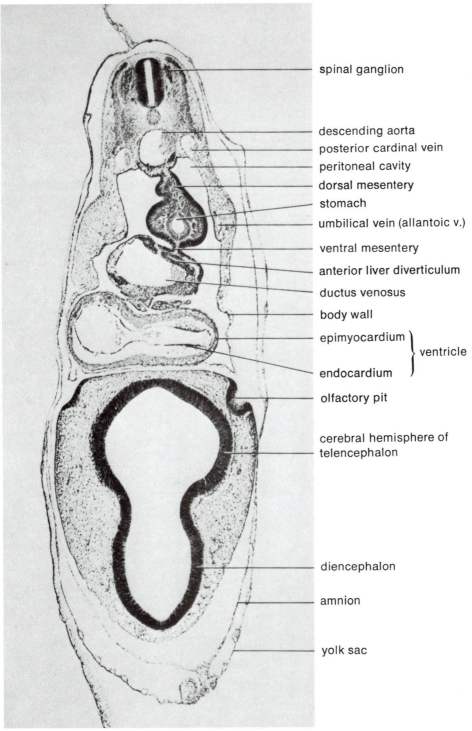

spinal ganglion

descending aorta
posterior cardinal vein
peritoneal cavity
dorsal mesentery
stomach
umbilical vein (allantoic v.)
ventral mesentery
anterior liver diverticulum
ductus venosus
body wall
epimyocardium ⎫
                ⎬ ventricle
endocardium   ⎭
olfactory pit

cerebral hemisphere of telencephalon

diencephalon

amnion

yolk sac

Figure 216 3-day chick embryo (stage 18), transverse section through olfactory pits (mag. 50X)

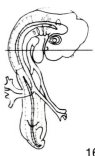

167

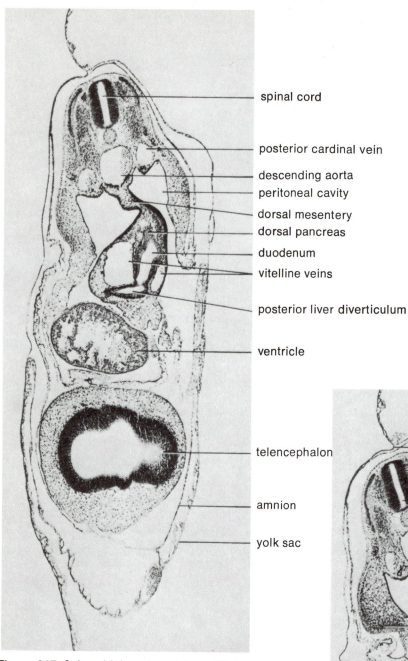

- spinal cord
- posterior cardinal vein
- descending aorta
- peritoneal cavity
- dorsal mesentery
- dorsal pancreas
- duodenum
- vitelline veins
- posterior liver diverticulum
- ventricle
- telencephalon
- amnion
- yolk sac

Figure 217 3-day chick embryo (stage 18), transverse section through liver and pancreas (mag. 50X)

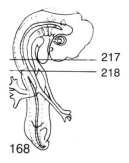

217
218

168

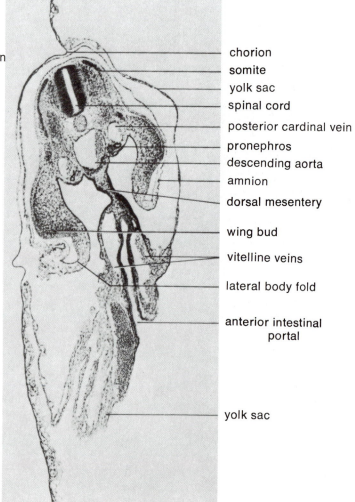

- chorion
- somite
- yolk sac
- spinal cord
- posterior cardinal vein
- pronephros
- descending aorta
- amnion
- dorsal mesentery
- wing bud
- vitelline veins
- lateral body fold
- anterior intestinal portal
- yolk sac

Figure 218 3-day chick embryo (stage 18), transverse section through anterior intestinal portal (mag. 50X)

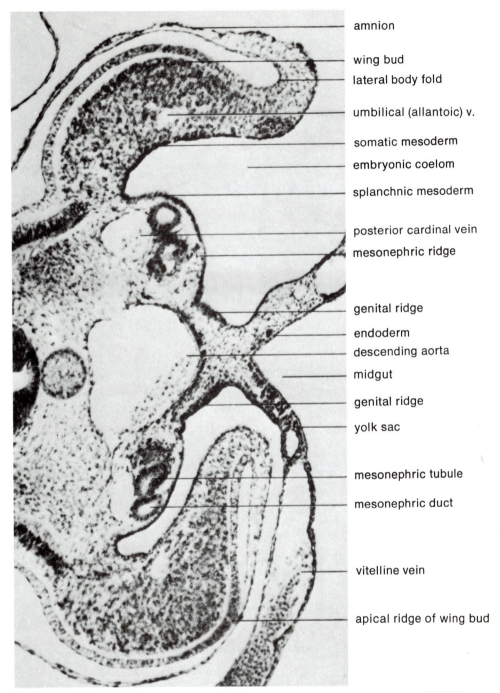

amnion

wing bud
lateral body fold

umbilical (allantoic) v.

somatic mesoderm
embryonic coelom
splanchnic mesoderm

posterior cardinal vein
mesonephric ridge

genital ridge
endoderm
descending aorta
midgut
genital ridge
yolk sac

mesonephric tubule

mesonephric duct

vitelline vein

apical ridge of wing bud

Figure 219 3-day chick embryo (stage 18), transverse section through genital ridge (mag. 140X)

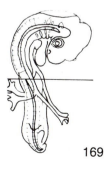

169

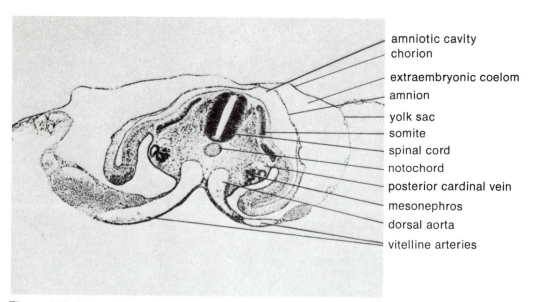

amniotic cavity
chorion

extraembryonic coelom
amnion

yolk sac
somite
spinal cord
notochord
posterior cardinal vein
mesonephros
dorsal aorta
vitelline arteries

Figure 220 3-day embryo (stage 18), transverse section through vitelline arteries (mag. 50X)

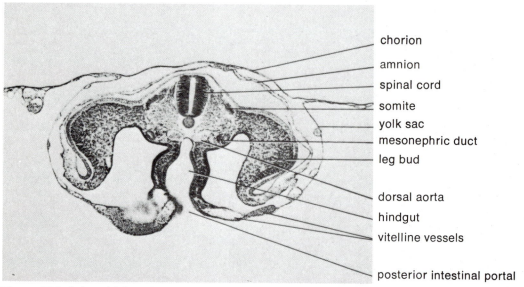

chorion

amnion

spinal cord

somite
yolk sac
mesonephric duct

leg bud

dorsal aorta

hindgut

vitelline vessels

posterior intestinal portal

Figure 221 3-day chick embryo (stage 18), transverse section through posterior intestinal portal (mag. 50X)

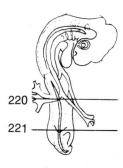

220

221

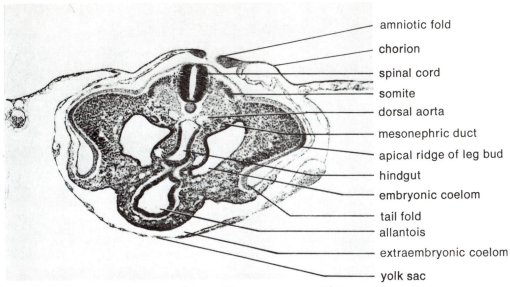

amniotic fold
chorion
spinal cord
somite
dorsal aorta
mesonephric duct
apical ridge of leg bud
hindgut
embryonic coelom
tail fold
allantois
extraembryonic coelom
yolk sac

Figure 222 3-day chick embryo (stage 18), transverse section through allantois (mag. 50X)

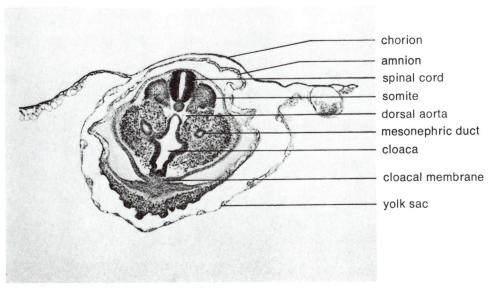

chorion
amnion
spinal cord
somite
dorsal aorta
mesonephric duct
cloaca
cloacal membrane
yolk sac

Figure 223 3-day chick embryo (stage 18), transverse section through cloaca (mag. 50X)

222
223

171

17. The 3½-Day Chick Embryo (Stage 20–21)

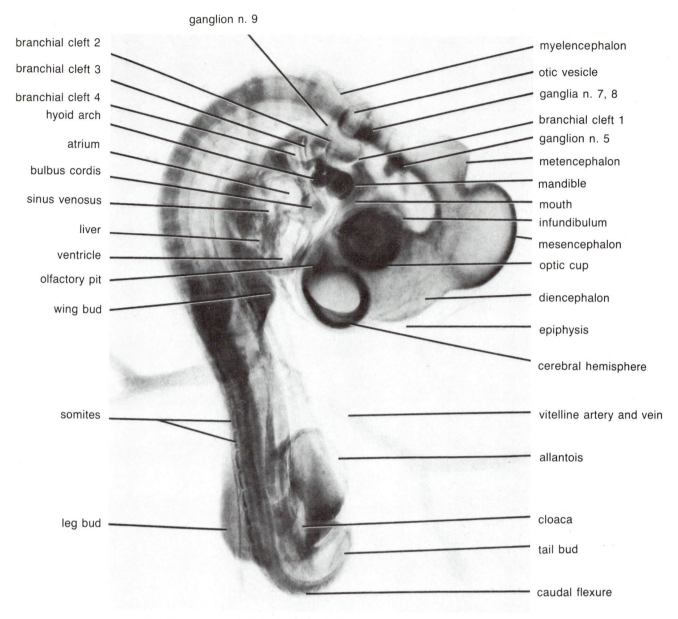

ganglion n. 9

branchial cleft 2
branchial cleft 3
branchial cleft 4
hyoid arch
atrium
bulbus cordis
sinus venosus
liver
ventricle
olfactory pit
wing bud

somites

leg bud

myelencephalon
otic vesicle
ganglia n. 7, 8
branchial cleft 1
ganglion n. 5
metencephalon
mandible
mouth
infundibulum
mesencephalon
optic cup

diencephalon

epiphysis

cerebral hemisphere

vitelline artery and vein

allantois

cloaca

tail bud

caudal flexure

Figure 224 3½-day chick embryo (stage 20), whole mount (mag. 20X)

174

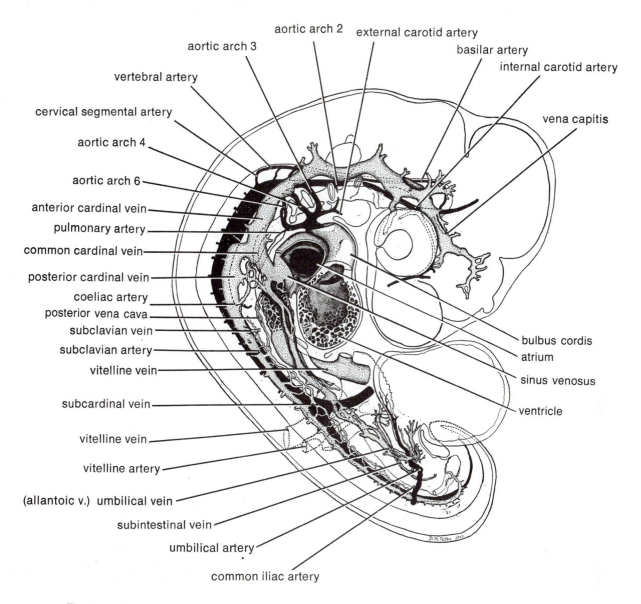

aortic arch 2
aortic arch 3
external carotid artery
basilar artery
internal carotid artery
vertebral artery
cervical segmental artery
vena capitis
aortic arch 4
aortic arch 6
anterior cardinal vein
pulmonary artery
common cardinal vein
posterior cardinal vein
coeliac artery
posterior vena cava
subclavian vein
subclavian artery
vitelline vein
subcardinal vein
vitelline vein
vitelline artery
(allantoic v.) umbilical vein
subintestinal vein
umbilical artery
common iliac artery
bulbus cordis
atrium
sinus venosus
ventricle

Figure 225

Reconstruction of circulatory system of 3½-day chick (stage 21) (mag. 18X). (From *Early Embryology of the Chick*, 4th ed., by Bradley M. Patten. Copyright 1951 by McGraw-Hill Book Co. Copyright renewed 1957 by B. M. Patten. Used with permission of the McGraw-Hill Book Co.)

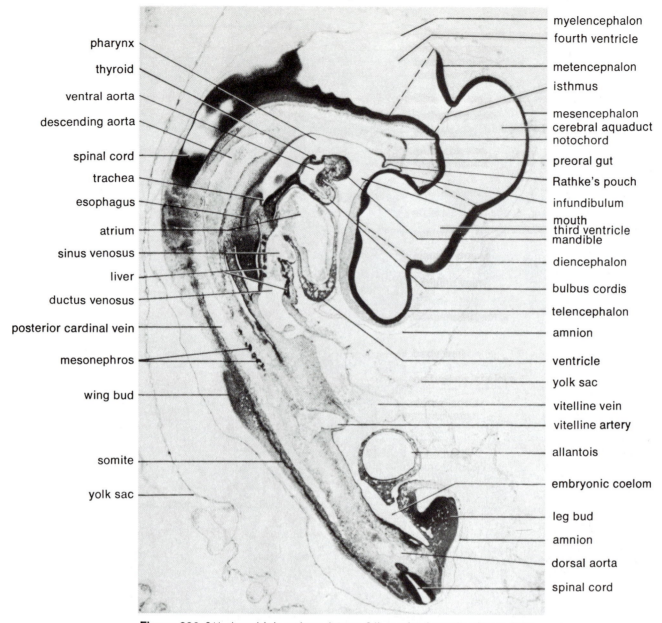

pharynx

thyroid

ventral aorta

descending aorta

spinal cord

trachea

esophagus

atrium

sinus venosus

liver

ductus venosus

posterior cardinal vein

mesonephros

wing bud

somite

yolk sac

myelencephalon

fourth ventricle

metencepnalon

isthmus

mesencephalon

cerebral aquaduct

notochord

preoral gut

Rathke's pouch

infundibulum

mouth

third ventricle

mandible

diencephalon

bulbus cordis

telencephalon

amnion

ventricle

yolk sac

vitelline vein

vitelline artery

allantois

embryonic coelom

leg bud

amnion

dorsal aorta

spinal cord

Figure 226 3½-day chick embryo (stage 21), sagittal section (mag. 25X)

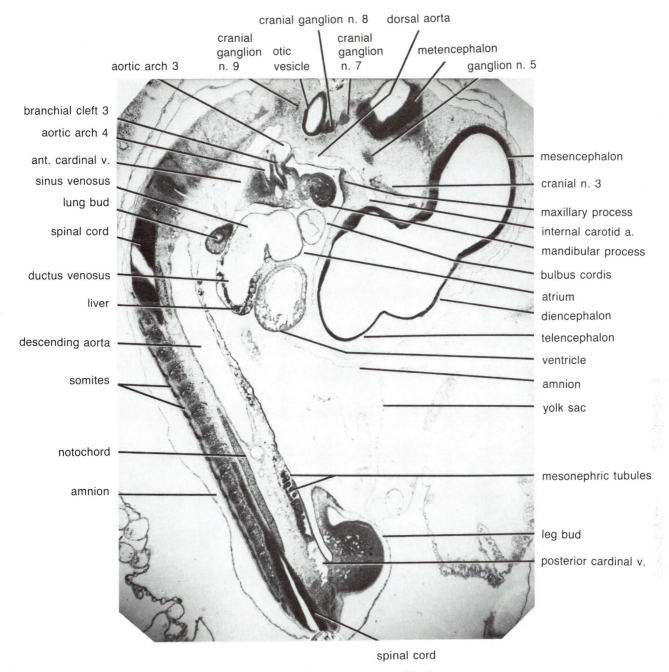

cranial ganglion n. 8   dorsal aorta

cranial ganglion n. 9   otic vesicle   cranial ganglion n. 7   metencephalon

aortic arch 3   ganglion n. 5

branchial cleft 3

aortic arch 4

ant. cardinal v.

sinus venosus

lung bud

spinal cord

ductus venosus

liver

descending aorta

somites

notochord

amnion

mesencephalon

cranial n. 3

maxillary process

internal carotid a.

mandibular process

bulbus cordis

atrium

diencephalon

telencephalon

ventricle

amnion

yolk sac

mesonephric tubules

leg bud

posterior cardinal v.

spinal cord

Figure 227   3½-day chick embryo (stage 21), parasagittal section (mag. 25 ×).

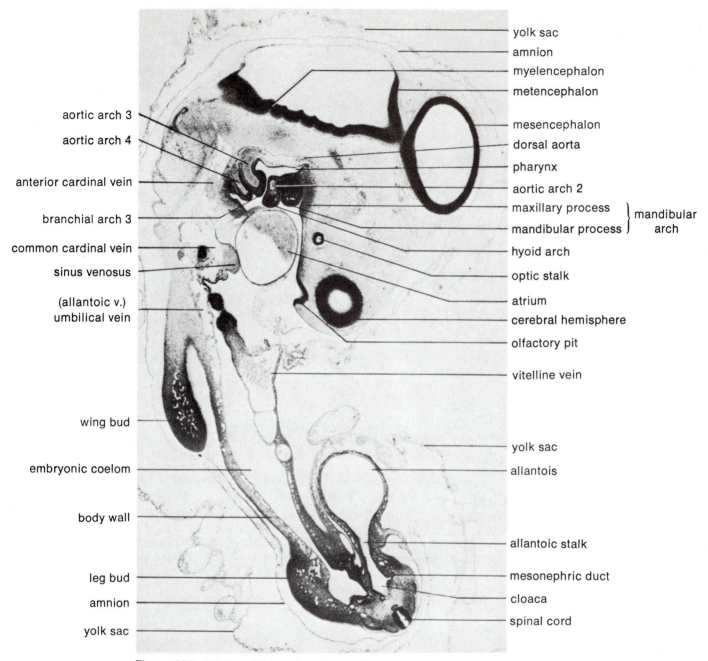

Figure 228 3½-day chick embryo (stage 21), parasagittal section (mag. 25X)

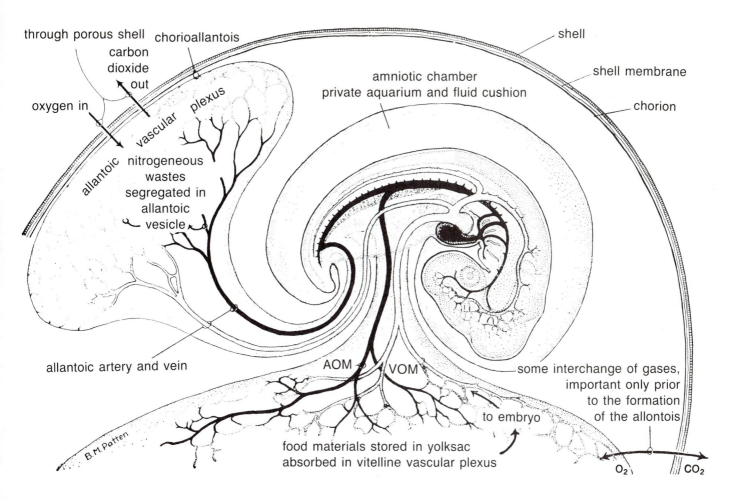

through porous shell
carbon dioxide out
oxygen in
chorioallantois
vascular plexus
allantoic
nitrogeneous wastes segregated in allantoic vesicle
allantoic artery and vein
amniotic chamber private aquarium and fluid cushion
shell
shell membrane
chorion
AOM
VOM
to embryo
food materials stored in yolksac absorbed in vitelline vascular plexus
some interchange of gases, important only prior to the formation of the allontois
O_2
CO_2
B.M.Patten

Figure 229 Schematic diagram showing the arrangement of main circulatory channels in a 4-day chick embryo. The sites of some of the extraembryonic interchanges important in its physiology are indicated by the labeling. The vessels within the embryo carry food and oxygen to all its growing tissues, and relieve them of the waste products incident to their metabolism. Abbreviations: AOM, omphalomesenteric (vitelline) artery; VOM, omphalomesenteric (vitelline) vein. [From B. M. Patten, The first heart beats and the beginning of embryonic circulation. *American Scientist* 39:225 (1951). Used with permission of *American Scientist*.]

18. The 5–6-mm Pig Embryo

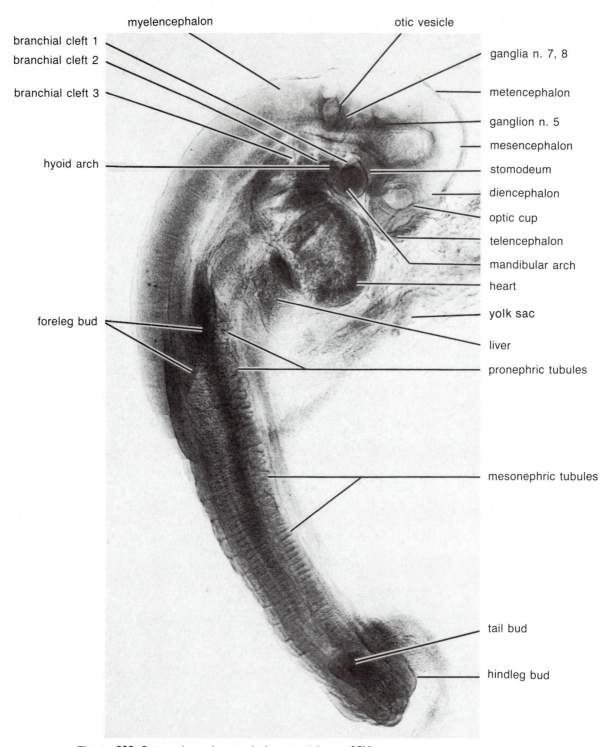

myelencephalon

otic vesicle

branchial cleft 1
branchial cleft 2

ganglia n. 7, 8

branchial cleft 3

metencephalon

ganglion n. 5

mesencephalon

hyoid arch

stomodeum

diencephalon

optic cup

telencephalon

mandibular arch

heart

foreleg bud

yolk sac

liver

pronephric tubules

mesonephric tubules

tail bud

hindleg bud

Figure 230 5-mm pig embryo, whole mount (mag. 25X)

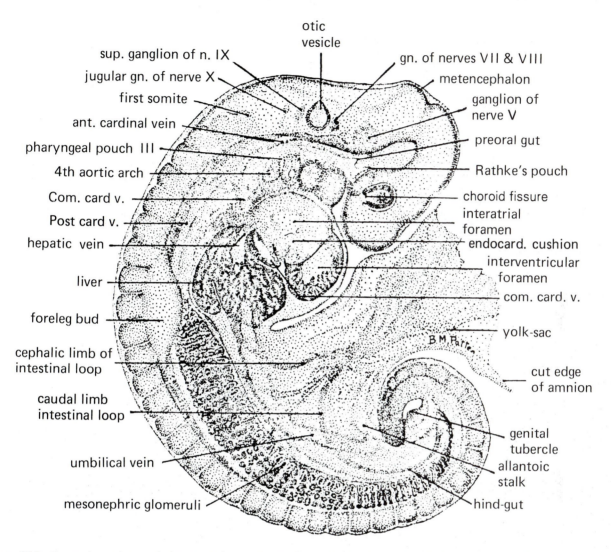

otic
vesicle

sup. ganglion of n. IX

jugular gn. of nerve X

first somite

ant. cardinal vein

pharyngeal pouch III

4th aortic arch

Com. card v.

Post card v.

hepatic vein

liver

foreleg bud

cephalic limb of
intestinal loop

caudal limb
intestinal loop

umbilical vein

mesonephric glomeruli

gn. of nerves VII & VIII

metencephalon

ganglion of
nerve V

preoral gut

Rathke's pouch

choroid fissure

interatrial
foramen

endocard. cushion

interventricular
foramen

com. card. v.

yolk-sac

cut edge
of amnion

genital
tubercle

allantoic
stalk

hind-gut

B M Patten

Figure 231 5mm pig embryo, whole mount (mag. 17 x). (From B. M. Patten, *The Embryology of the Pig*. Copyright 1948 by the McGraw-Hill Book Co. Copyright renewed 1959 by B. M. Patten. Used with permission of the McGraw-Hill Book Co.)

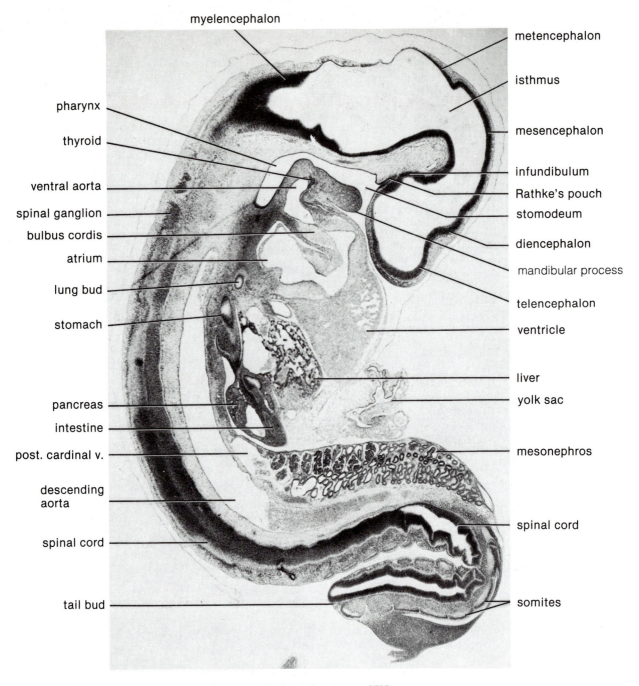

myelencephalon

metencephalon

isthmus

mesencephalon

pharynx

thyroid

ventral aorta

spinal ganglion

bulbus cordis

atrium

lung bud

stomach

pancreas

intestine

post. cardinal v.

descending aorta

spinal cord

tail bud

infundibulum

Rathke's pouch

stomodeum

diencephalon

mandibular process

telencephalon

ventricle

liver

yolk sac

mesonephros

spinal cord

somites

Figure 232 6-mm pig embryo, sagittal section (mag. 25X)

19. The 10-mm Pig Embryo

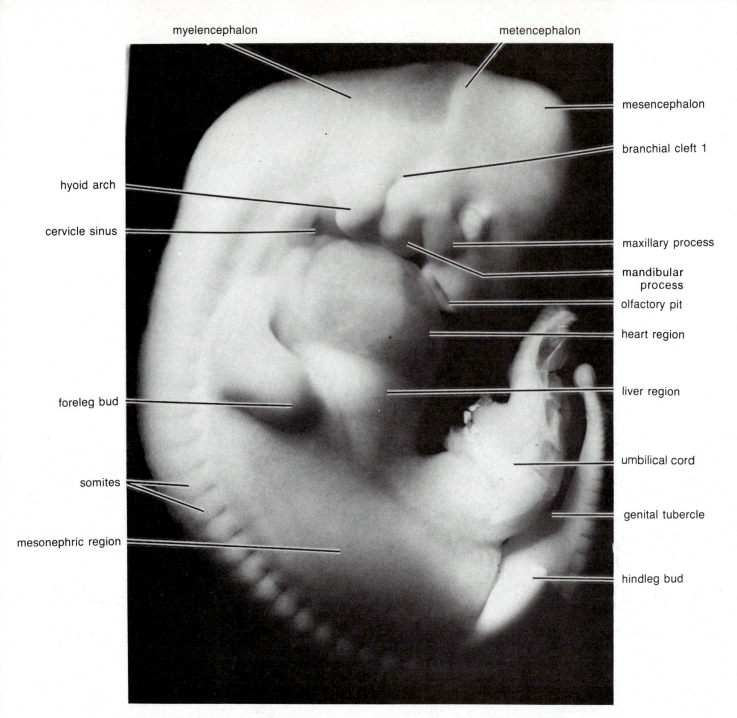

myelencephalon

metencephalon

mesencephalon

branchial cleft 1

hyoid arch

cervicle sinus

maxillary process

mandibular
process

olfactory pit

heart region

foreleg bud

liver region

umbilical cord

somites

genital tubercle

mesonephric region

hindleg bud

Figure 233 10-mm pig embryo, opaque mount (mag. 17X; incident illumination)

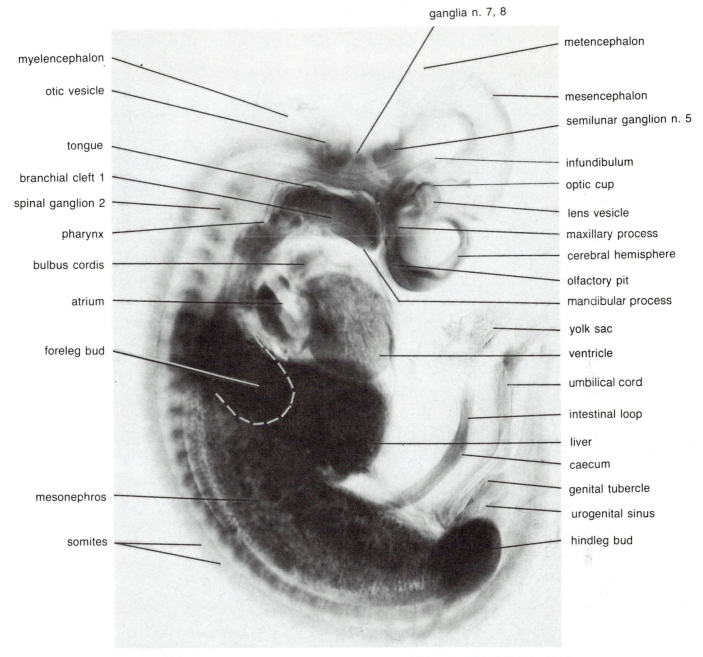

ganglia n. 7, 8

metencephalon

myelencephalon

mesencephalon

otic vesicle

semilunar ganglion n. 5

tongue

infundibulum

branchial cleft 1

optic cup

spinal ganglion 2

lens vesicle

pharynx

maxillary process

bulbus cordis

cerebral hemisphere

atrium

olfactory pit

mandibular process

foreleg bud

yolk sac

ventricle

umbilical cord

intestinal loop

liver

caecum

mesonephros

genital tubercle

urogenital sinus

somites

hindleg bud

Figure 234 10-mm pig embryo, whole mount (mag. 16X; transmitted illumination)

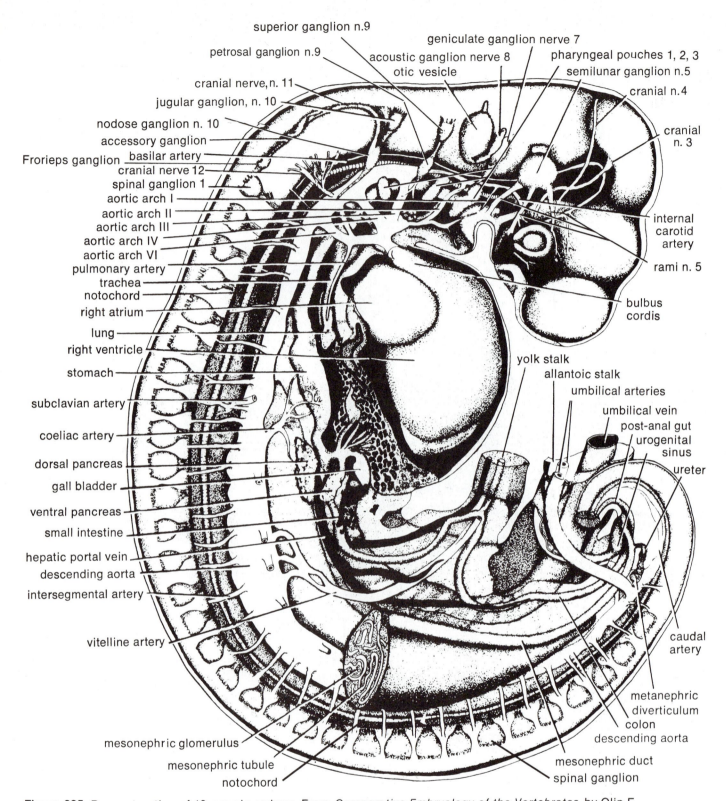

Figure 235 Reconstruction of 10-mm pig embryo. From *Comparative Embryology of the Vertebrates* by Olin E. Nelson. Copyright 1953 by the Blakiston Company, Inc. Used with permission of McGraw-Hill Book Company.)

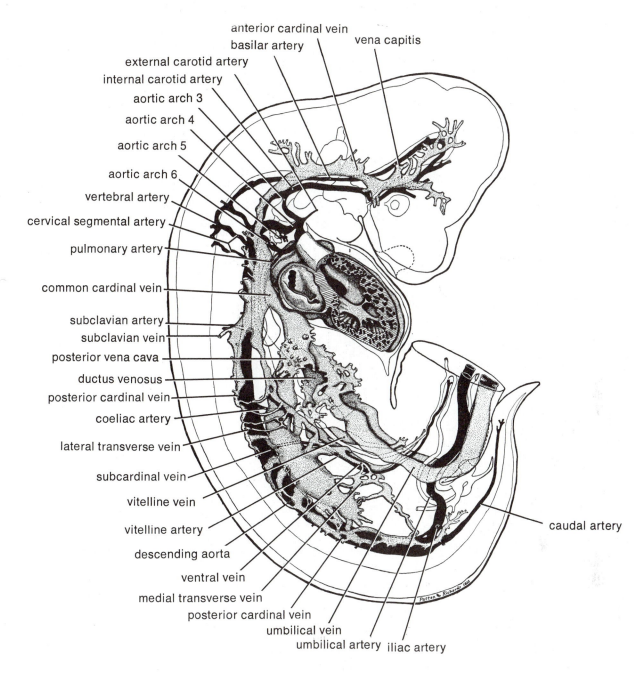

anterior cardinal vein
basilar artery
vena capitis
external carotid artery
internal carotid artery
aortic arch 3
aortic arch 4
aortic arch 5
aortic arch 6
vertebral artery
cervical segmental artery
pulmonary artery
common cardinal vein
subclavian artery
subclavian vein
posterior vena cava
ductus venosus
posterior cardinal vein
coeliac artery
lateral transverse vein
subcardinal vein
vitelline vein
vitelline artery
descending aorta
ventral vein
medial transverse vein
posterior cardinal vein
umbilical vein
umbilical artery
iliac artery
caudal artery

Figure 236

Reconstruction of the circulatory system of a 9.4-mm pig embryo (mag. 14X). (From *Embryology of the Pig,* 3rd ed., *by* Bradley M. Patten. Copyright 1948 by McGraw-Hill Book Co. Copyright renewed 1959 by B. M. Patten. Used with permission of the McGraw-Hill Book Co.)

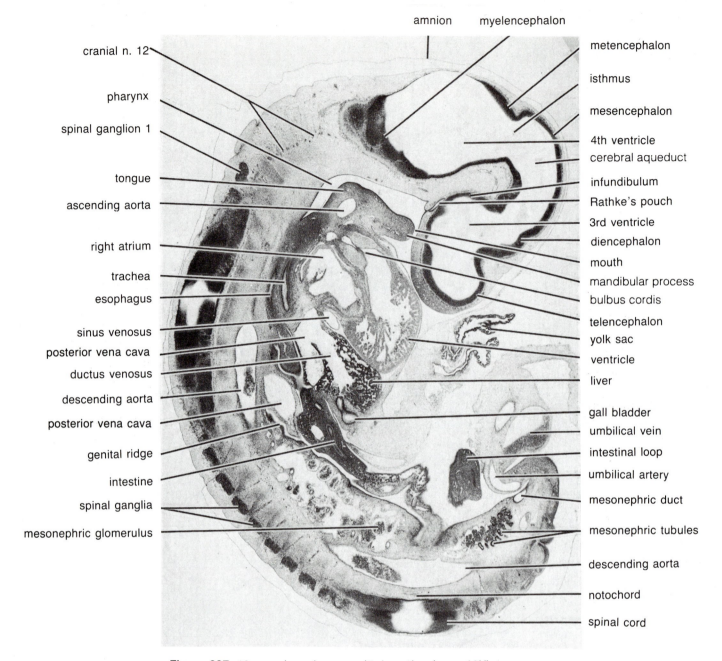

amnion   myelencephalon

cranial n. 12

pharynx

spinal ganglion 1

tongue

ascending aorta

right atrium

trachea

esophagus

sinus venosus

posterior vena cava

ductus venosus

descending aorta

posterior vena cava

genital ridge

intestine

spinal ganglia

mesonephric glomerulus

metencephalon

isthmus

mesencephalon

4th ventricle
cerebral aqueduct

infundibulum

Rathke's pouch

3rd ventricle

diencephalon

mouth

mandibular process

bulbus cordis

telencephalon

yolk sac

ventricle

liver

gall bladder

umbilical vein

intestinal loop

umbilical artery

mesonephric duct

mesonephric tubules

descending aorta

notochord

spinal cord

Figure 237 10-mm pig embryo, sagittal section (mag. 16X)

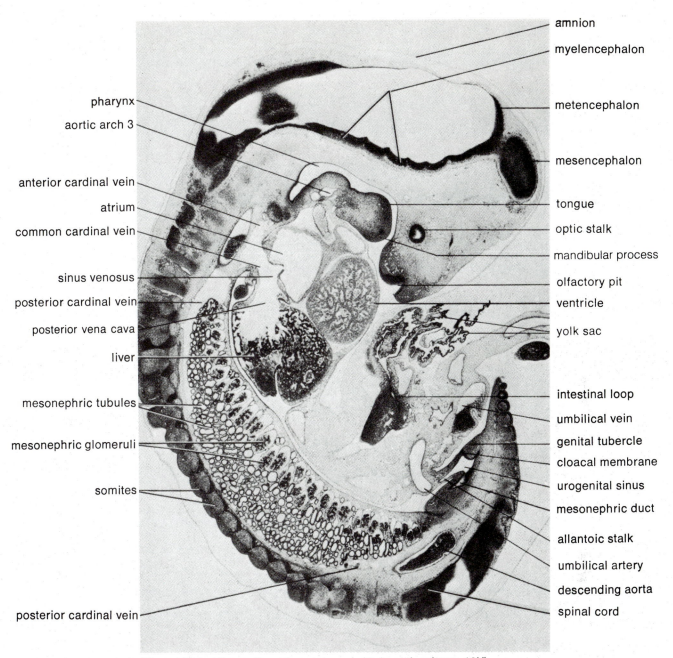

pharynx
aortic arch 3

anterior cardinal vein
atrium
common cardinal vein

sinus venosus
posterior cardinal vein
posterior vena cava
liver

mesonephric tubules

mesonephric glomeruli

somites

posterior cardinal vein

amnion
myelencephalon

metencephalon

mesencephalon

tongue
optic stalk
mandibular process
olfactory pit
ventricle
yolk sac

intestinal loop
umbilical vein
genital tubercle
cloacal membrane
urogenital sinus
mesonephric duct
allantoic stalk
umbilical artery
descending aorta
spinal cord

Figure 238 10-mm pig embryo, parasagittal section (mag. 16X)

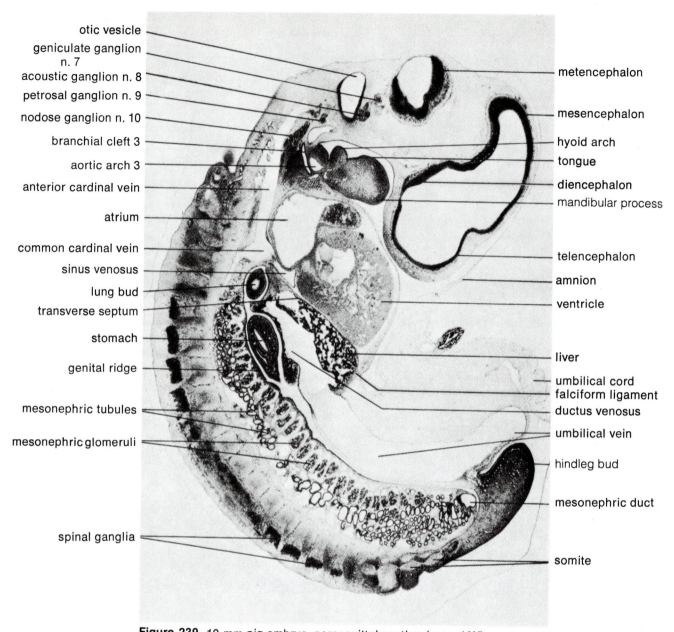

otic vesicle

geniculate ganglion n. 7

acoustic ganglion n. 8

petrosal ganglion n. 9

nodose ganglion n. 10

branchial cleft 3

aortic arch 3

anterior cardinal vein

atrium

common cardinal vein

sinus venosus

lung bud

transverse septum

stomach

genital ridge

mesonephric tubules

mesonephric glomeruli

spinal ganglia

metencephalon

mesencephalon

hyoid arch

tongue

diencephalon

mandibular process

telencephalon

amnion

ventricle

liver

umbilical cord
falciform ligament

ductus venosus

umbilical vein

hindleg bud

mesonephric duct

somite

Figure 239 10-mm pig embryo, parasagittal section (mag. 16X)

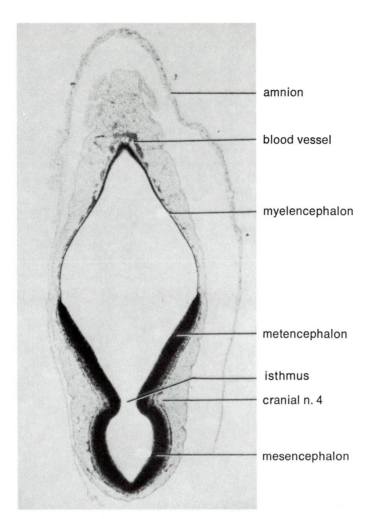

amnion

blood vessel

myelencephalon

metencephalon

isthmus

cranial n. 4

mesencephalon

Figure 240 10-mm pig embryo, transverse section through cranial nerve 4 (mag. 30X)

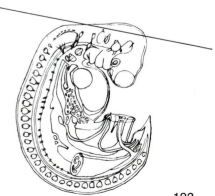

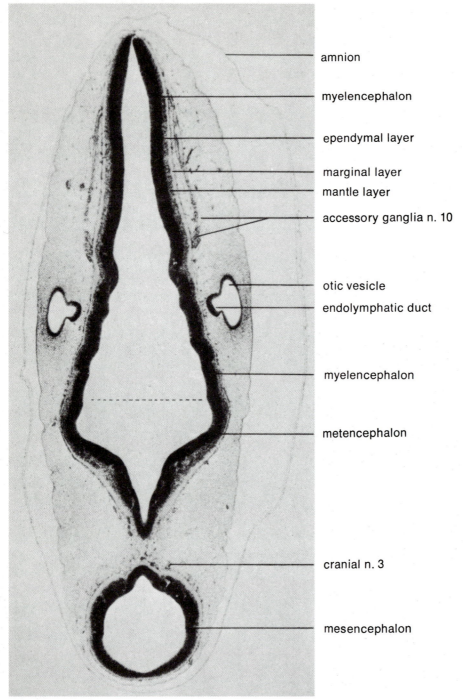

amnion

myelencephalon

ependymal layer

marginal layer

mantle layer

accessory ganglia n. 10

otic vesicle

endolymphatic duct

myelencephalon

metencephalon

cranial n. 3

mesencephalon

Figure 241 10-mm pig embryo, transverse section through accessory cranial ganglia (mag. 30X)

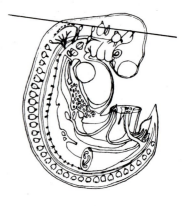

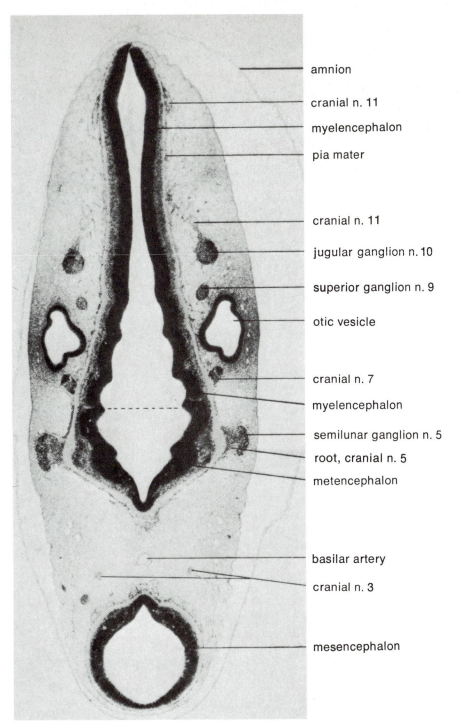

amnion

cranial n. 11

myelencephalon

pia mater

cranial n. 11

jugular ganglion n. 10

superior ganglion n. 9

otic vesicle

cranial n. 7

myelencephalon

semilunar ganglion n. 5

root, cranial n. 5

metencephalon

basilar artery

cranial n. 3

mesencephalon

Figure 242 10-mm pig embryo, transverse section through
jugular and superior cranial ganglia (mag. 30X)

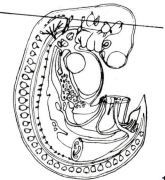

195

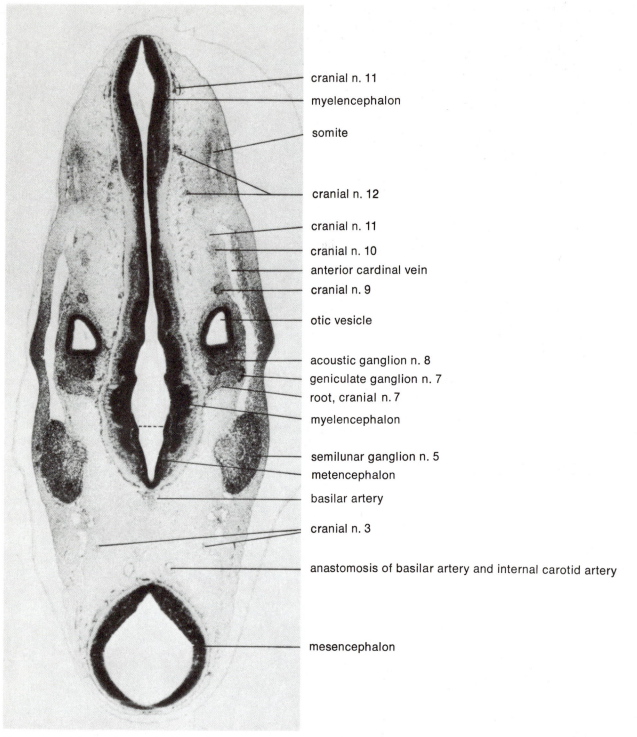

cranial n. 11

myelencephalon

somite

cranial n. 12

cranial n. 11

cranial n. 10

anterior cardinal vein

cranial n. 9

otic vesicle

acoustic ganglion n. 8

geniculate ganglion n. 7

root, cranial n. 7

myelencephalon

semilunar ganglion n. 5

metencephalon

basilar artery

cranial n. 3

anastomosis of basilar artery and internal carotid artery

mesencephalon

Figure 243 10-mm pig embryo, transverse section through semilunar and geniculate cranial ganglia (mag. 30X)

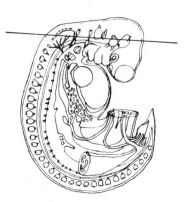

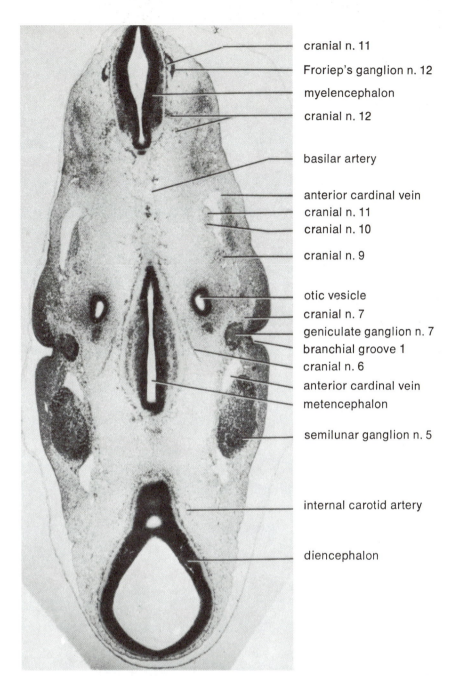

cranial n. 11

Froriep's ganglion n. 12

myelencephalon

cranial n. 12

basilar artery

anterior cardinal vein
cranial n. 11
cranial n. 10

cranial n. 9

otic vesicle
cranial n. 7
geniculate ganglion n. 7
branchial groove 1
cranial n. 6
anterior cardinal vein
metencephalon

semilunar ganglion n. 5

internal carotid artery

diencephalon

Figure 244 10-mm pig embryo, transverse section through cranial nerve 6 (mag. 30 X)

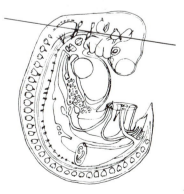

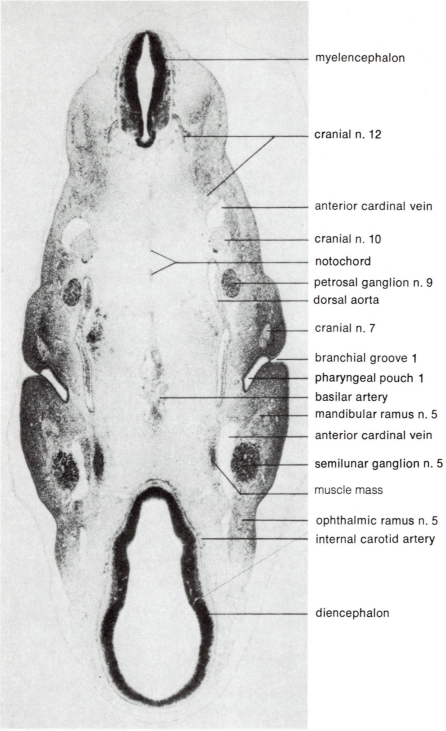

Figure 245 10-mm pig embryo, transverse section through pharyngeal pouch 1 (mag. 30X)

myelencephalon

cranial n. 12

anterior cardinal vein

cranial n. 10

notochord

petrosal ganglion n. 9

dorsal aorta

cranial n. 7

branchial groove 1

pharyngeal pouch 1

basilar artery

mandibular ramus n. 5

anterior cardinal vein

semilunar ganglion n. 5

muscle mass

ophthalmic ramus n. 5

internal carotid artery

diencephalon

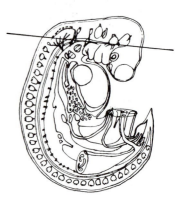

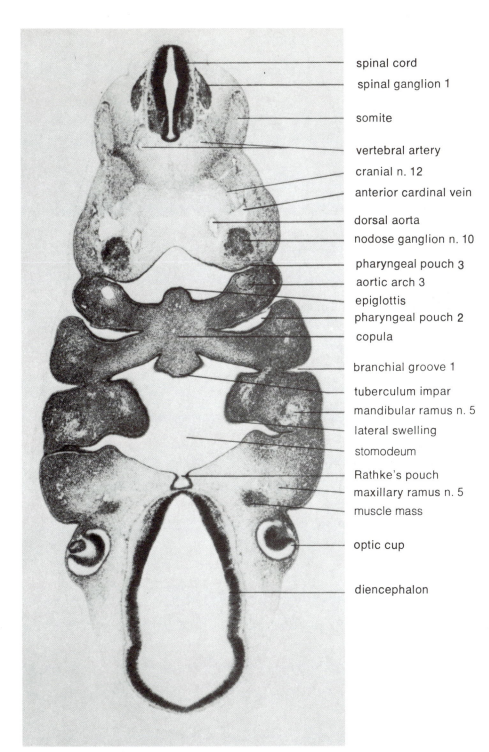

spinal cord

spinal ganglion 1

somite

vertebral artery

cranial n. 12

anterior cardinal vein

dorsal aorta

nodose ganglion n. 10

pharyngeal pouch 3

aortic arch 3

epiglottis

pharyngeal pouch 2

copula

branchial groove 1

tuberculum impar

mandibular ramus n. 5

lateral swelling

stomodeum

Rathke's pouch

maxillary ramus n. 5

muscle mass

optic cup

diencephalon

Figure 246 10-mm pig embryo, transverse section through Rathke's pouch (mag. 30X)

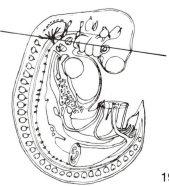

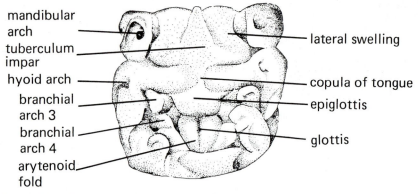

mandibular arch

tuberculum impar

hyoid arch

branchial arch 3

branchial arch 4

arytenoid fold

lateral swelling

copula of tongue

epiglottis

glottis

Figure 247   Floor of mouth and parynx of 10mm pig embryo after being cut away from upper head. (From L. B. Arey, *Developmental Anatomy*. Copyright 1965 by W. B. Saunders Co. Used with permission of the W. B. Saunders Co.)

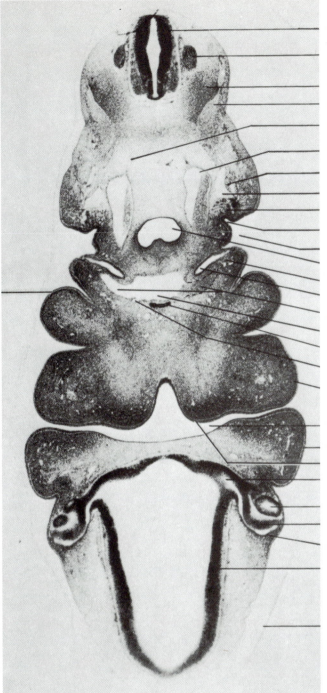

spinal cord

spinal ganglion 2

sclerotome

myotome

intersegmental artery

left dorsal aorta

cranial n. 12

anterior cardinal vein

cranial n. 10

branchial groove 4

cervical sinus

pharynx

parathyroid

pharyngeal pouch 3

aortic arch 3

ventral aorta

thyroid

external carotid artery

mandibular process

stomodeum

maxillary process

lateral swelling

optic stalk

lens vesicle

nervous layer of optic cup ⎫
                            ⎬ retina
pigmented layer of optic cup ⎭

diencephalon

amnion

hyoid arch

Figure 248 10-mm pig embryo, transverse section through thyroid (mag. 30X)

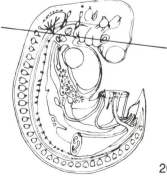

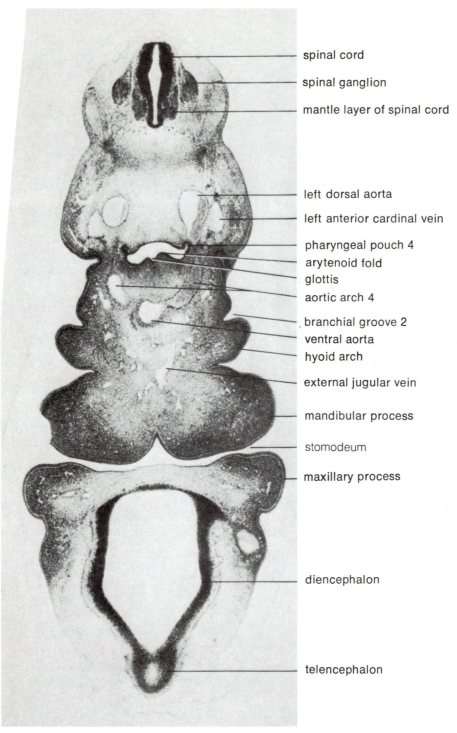

spinal cord

spinal ganglion

mantle layer of spinal cord

left dorsal aorta

left anterior cardinal vein

pharyngeal pouch 4

arytenoid fold

glottis

aortic arch 4

branchial groove 2

ventral aorta

hyoid arch

external jugular vein

mandibular process

stomodeum

maxillary process

diencephalon

telencephalon

Figure 249 10-mm pig embryo, transverse section through fourth aortic arch (mag. 30X)

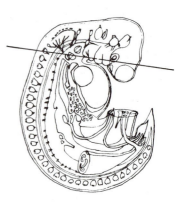

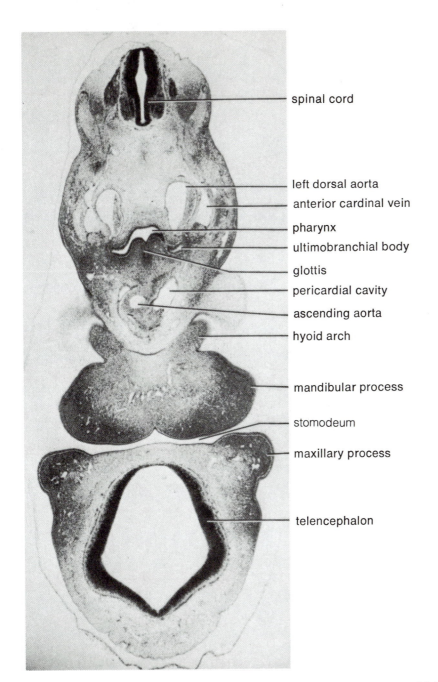

spinal cord

left dorsal aorta

anterior cardinal vein

pharynx

ultimobranchial body

glottis

pericardial cavity

ascending aorta

hyoid arch

mandibular process

stomodeum

maxillary process

telencephalon

Figure 250 10-mm pig embryo, transverse section through ultimobranchial body (mag. 30X)

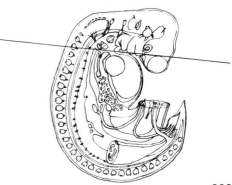

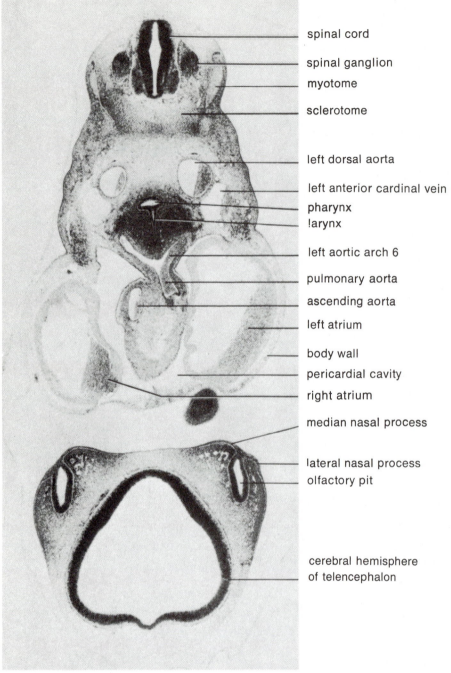

spinal cord

spinal ganglion

myotome

sclerotome

left dorsal aorta

left anterior cardinal vein

pharynx

larynx

left aortic arch 6

pulmonary aorta

ascending aorta

left atrium

body wall

pericardial cavity

right atrium

median nasal process

lateral nasal process

olfactory pit

cerebral hemisphere
of telencephalon

Figure 251 10-mm pig embryo, transverse section through
pulmonary aorta (mag. 30X)

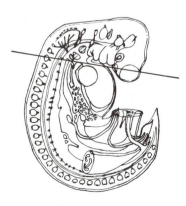

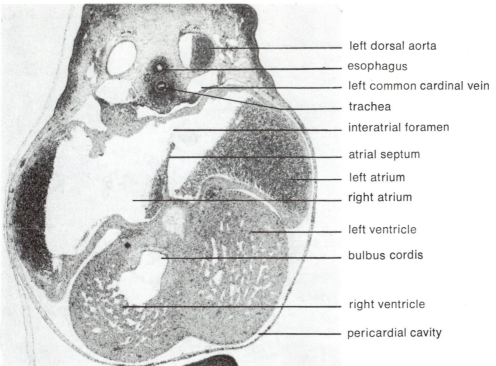

left dorsal aorta
esophagus
left common cardinal vein
trachea
interatrial foramen
atrial septum
left atrium
right atrium
left ventricle
bulbus cordis
right ventricle
pericardial cavity

Figure 252 10-mm pig embryo, transverse section through interatrial foramen (mag. 30X)

Figure 253 10-mm pig embryo, transverse section through interventricular foramen (mag. 30X)

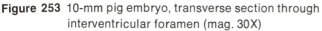

252
253

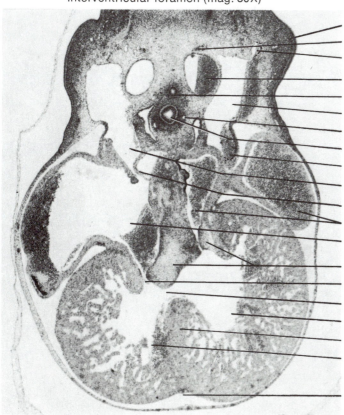

foreleg bud
sympathetic ganglion
subclavian vein
left dorsal aorta
esophagus
left common cardinal vein
left pulmonary artery
trachea
left horn, sinus venosus
right horn, sinus venosus
valvulae venosa
left atrium
right atrium
endocardial cushion
right atrioventricular canal
interventricular foramen
left ventricle
ventricular septum
right ventricle
interventricular sulcus

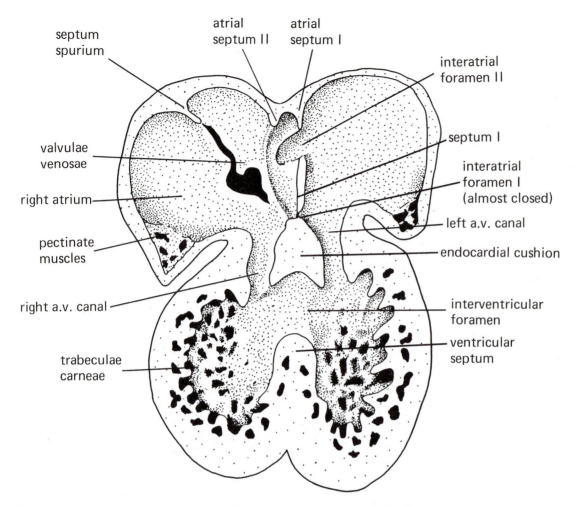

septum
spurium

atrial
septum II

atrial
septum I

interatrial
foramen II

valvulae
venosae

septum I

interatrial
foramen I
(almost closed)

right atrium

left a.v. canal

pectinate
muscles

endocardial cushion

right a.v. canal

interventricular
foramen

ventricular
septum

trabeculae
carneae

Figure 254 Reconstruction of the heart of a 9.4mm pig embryo; dorsal half of heart, interior view, near-frontal plane of section. (From B. M. Patten, *Embryology of the Pig*. Copyright 1948 by the McGraw-Hill Book Co. Copyright renewed 1959 by B. M. Patten. Used with permission of the McGraw-Hill Book Co.)

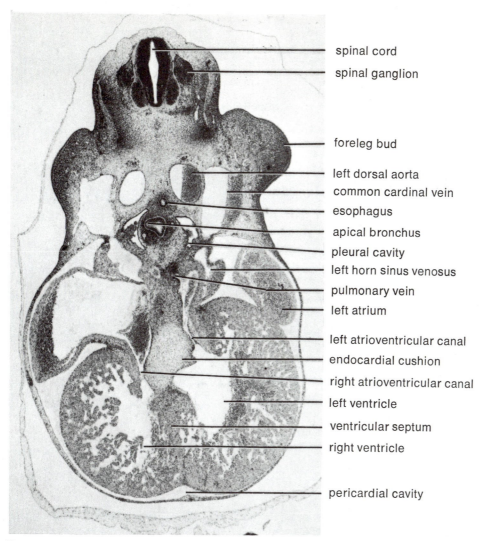

spinal cord

spinal ganglion

foreleg bud

left dorsal aorta

common cardinal vein

esophagus

apical bronchus

pleural cavity

left horn sinus venosus

pulmonary vein

left atrium

left atrioventricular canal

endocardial cushion

right atrioventricular canal

left ventricle

ventricular septum

right ventricle

pericardial cavity

Figure 255 10-mm pig embryo, transverse section through apical bronchus (mag. 30 X)

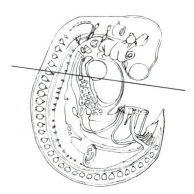

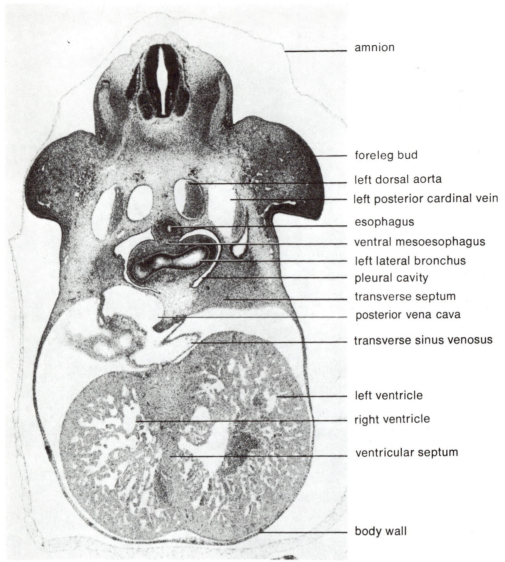

— amnion

— foreleg bud

— left dorsal aorta
— left posterior cardinal vein

— esophagus
— ventral mesoesophagus
— left lateral bronchus
— pleural cavity
— transverse septum
— posterior vena cava

— transverse sinus venosus

— left ventricle

— right ventricle

— ventricular septum

— body wall

Figure 256 10-mm pig embryo, transverse section through lateral lung buds (mag. 30X)

Figure 257 10-mm pig embryo, transverse section through caudal lung buds (mag. 30X)

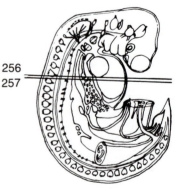

256
257

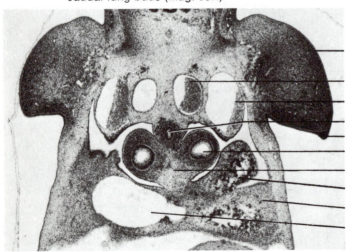

— foreleg bud

— left dorsal aorta
— left posterior cardinal vein
— esophagus
— apical ridge
— left stem bronchus
— ventral mesoesophagus
— liver
— transverse septum
— posterior vena cava

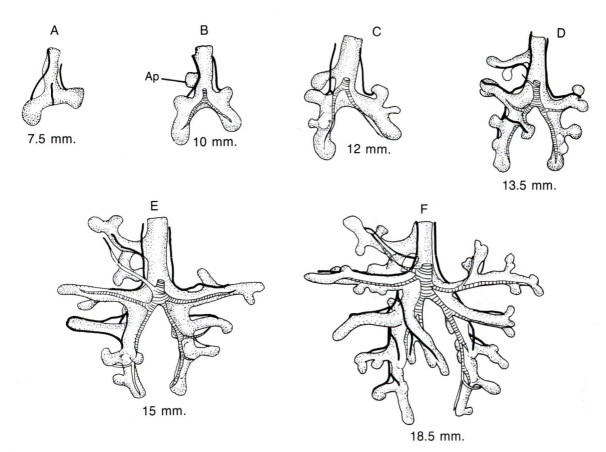

A 7.5 mm.

B Ap 10 mm.

C 12 mm.

D 13.5 mm.

E 15 mm.

F 18.5 mm.

Figure 258 Development of the trachea, bronchi, and lungs in the pig embryo. Pulmonary arteries are black; veins are cross hatched. Ap, apical bronchus. Compare 10mm pig sections to B (10mm). (From B. M. Patten, *Embryology of the Pig*. Copyright 1948 by the McGraw-Hill Book Co. Copyright renewed 1959 by B. M. Patten. Used with permission of the McGraw-Hill Book Co.)

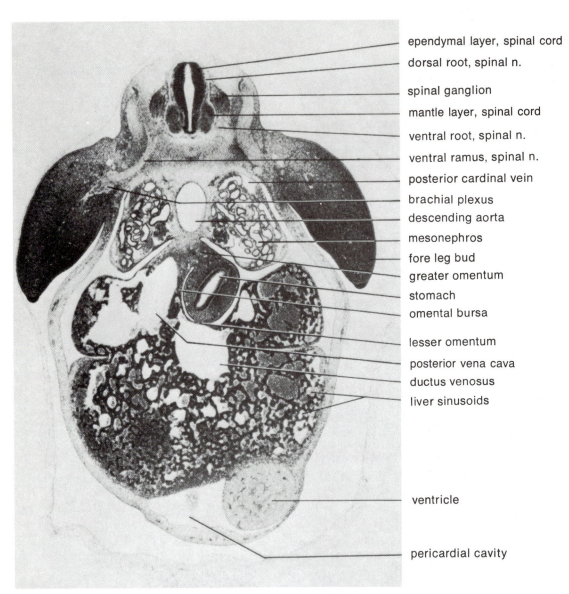

ependymal layer, spinal cord

dorsal root, spinal n.

spinal ganglion

mantle layer, spinal cord

ventral root, spinal n.

ventral ramus, spinal n.

posterior cardinal vein

brachial plexus

descending aorta

mesonephros

fore leg bud

greater omentum

stomach

omental bursa

lesser omentum

posterior vena cava

ductus venosus

liver sinusoids

ventricle

pericardial cavity

Figure 259 10-mm pig embryo, transverse section through ductus venosus
(mag. 30X)

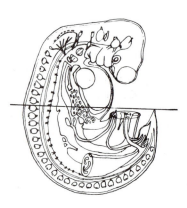

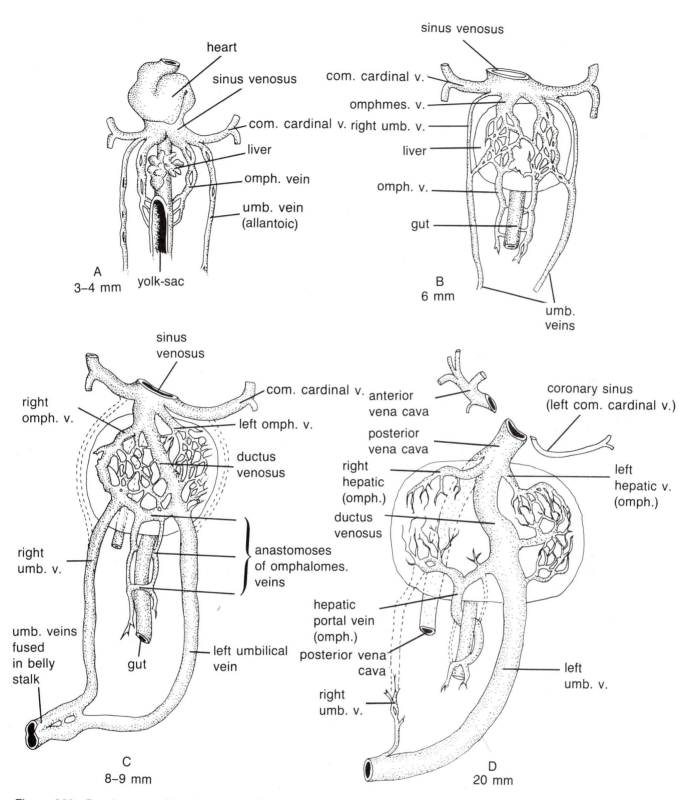

Figure 260 Development of the hepatic portal circulation and the umbilical veins in the pig embryo. Compare 10mm pig section to C (8–9mm). (From B. M. Patten, *Embryology of the Pig*. Copyright 1948 by the McGraw-Hill Book Co. Copyright renewed 1959 by B. M. Patten. Used with permission of the McGraw-Hill Book Co.)

ependymal layer
of spinal cord

marginal layer of spinal cord

somite

spinal ganglion

mantle layer of spinal cord

ventral root,
spinal nerve

posterior cardinal v.

peritoneal cavity

mesonephric tubule

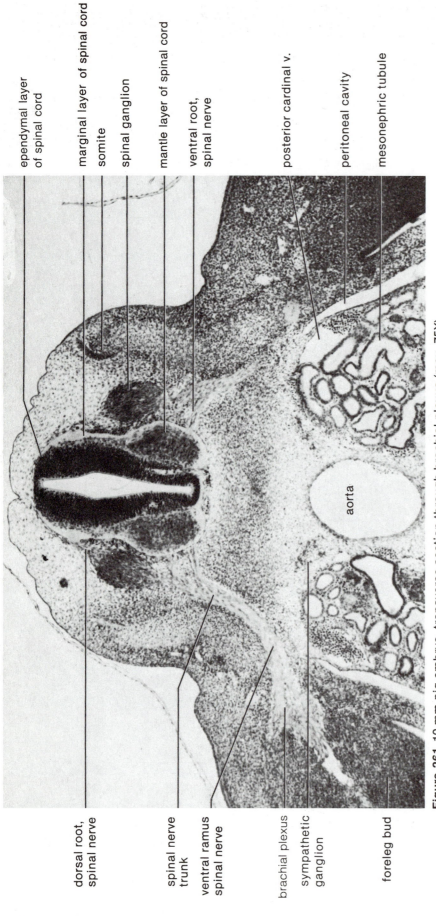

aorta

dorsal root,
spinal nerve

spinal nerve
trunk

ventral ramus
spinal nerve

brachial plexus

sympathetic
ganglion

foreleg bud

Figure 261 10-mm pig embryo, transverse section through brachial plexus (max. 75X)

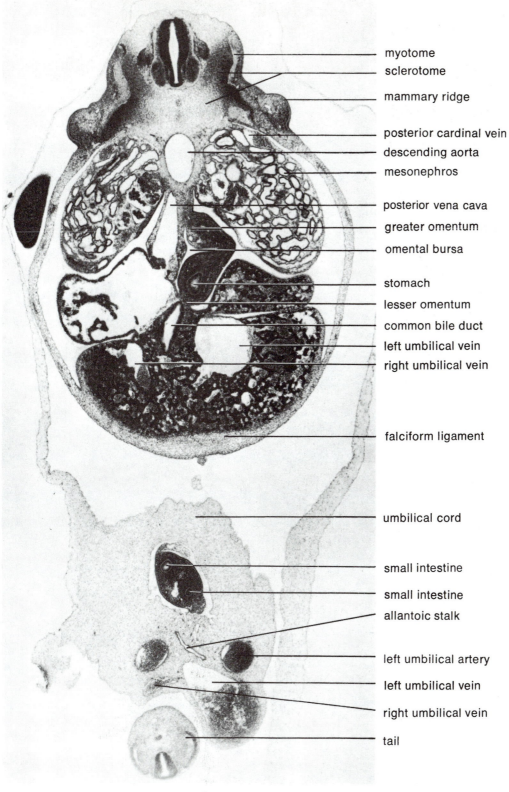

myotome

sclerotome

mammary ridge

posterior cardinal vein

descending aorta

mesonephros

posterior vena cava

greater omentum

omental bursa

stomach

lesser omentum

common bile duct

left umbilical vein

right umbilical vein

falciform ligament

umbilical cord

small intestine

small intestine

allantoic stalk

left umbilical artery

left umbilical vein

right umbilical vein

tail

Figure 262 10-mm pig embryo, transverse section through common bile duct
(mag. 30X)

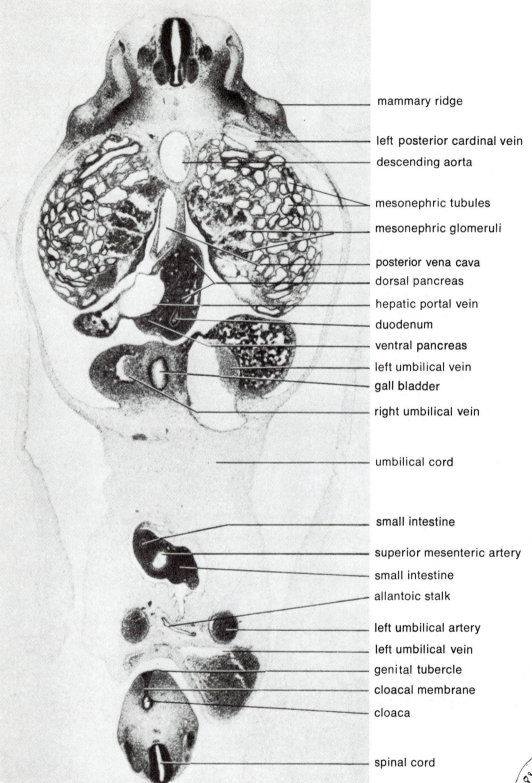

mammary ridge

left posterior cardinal vein
descending aorta

mesonephric tubules
mesonephric glomeruli

posterior vena cava
dorsal pancreas
hepatic portal vein
duodenum
ventral pancreas
left umbilical vein
gall bladder
right umbilical vein

umbilical cord

small intestine

superior mesenteric artery
small intestine
allantoic stalk

left umbilical artery
left umbilical vein
genital tubercle
cloacal membrane
cloaca

spinal cord

Figure 263 10-mm pig embryo, transverse section through gall bladder (mag. 30X)

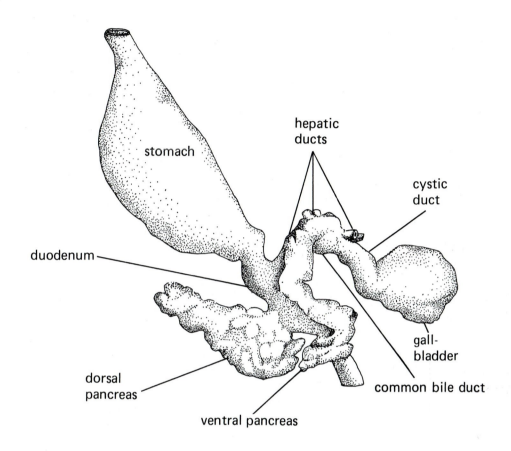

Figure 264 Reconstruction of the stomach and duodenum of a 9.4mm pig embryo showing the rudiments of the pancreas and liver. (From B. M. Patten, *Embryology of the Pig*. Copyright 1948 by the McGraw-Hill Book Co. Copyright renewed 1959 by B. M. Patten. Used with permission of the McGraw-Hill Book Co.)

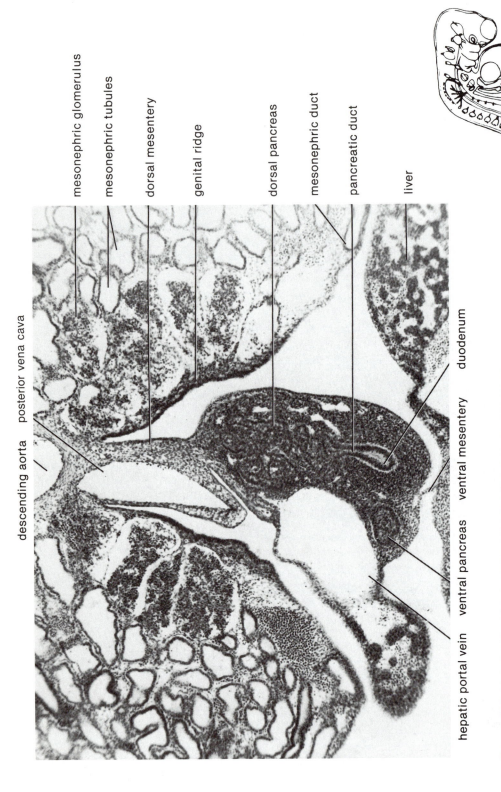

mesonephric glomerulus

mesonephric tubules

dorsal mesentery

genital ridge

dorsal pancreas

mesonephric duct

pancreatic duct

liver

descending aorta

posterior vena cava

duodenum

ventral mesentery

hepatic portal vein

ventral pancreas

Figure 265 10-mm pig embryo, transverse section through pancreas (mag. 75X)

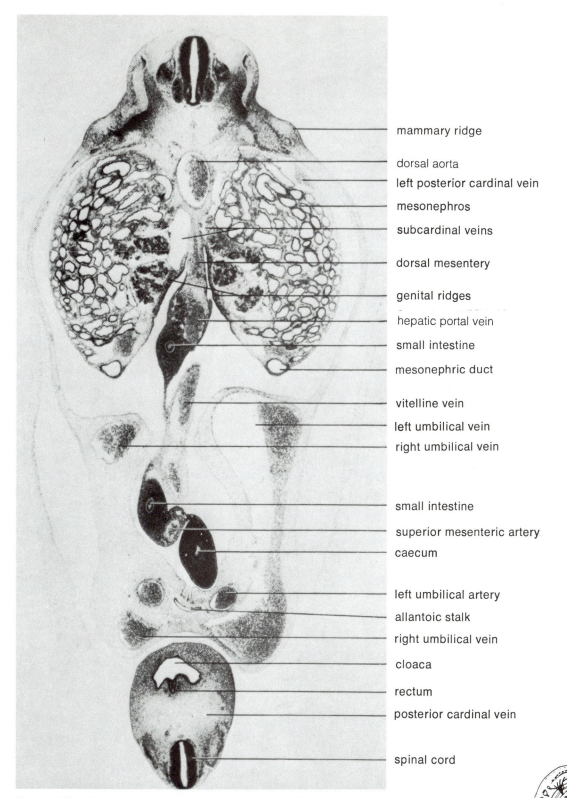

mammary ridge

dorsal aorta
left posterior cardinal vein
mesonephros
subcardinal veins

dorsal mesentery

genital ridges
hepatic portal vein
small intestine
mesonephric duct

vitelline vein
left umbilical vein
right umbilical vein

small intestine
superior mesenteric artery
caecum

left umbilical artery
allantoic stalk
right umbilical vein
cloaca
rectum
posterior cardinal vein

spinal cord

Figure 266 10-mm pig embryo, transverse section through genital ridge
(mag. 30X)

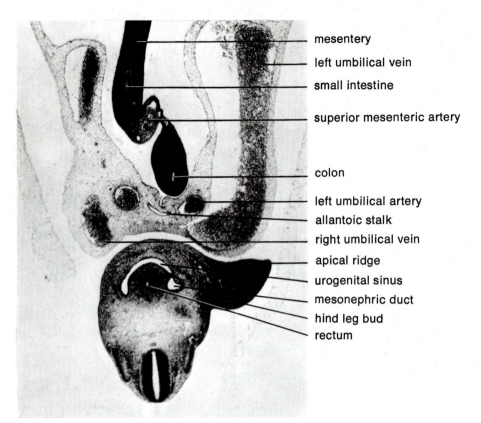

mesentery

left umbilical vein

small intestine

superior mesenteric artery

colon

left umbilical artery

allantoic stalk

right umbilical vein

apical ridge

urogenital sinus

mesonephric duct

hind leg bud

rectum

Figure 267 10-mm pig embryo, transverse section through urogenital sinus (mag. 30X)

Figure 268 10-mm pig embryo, transverse section through metanephros (mag. 30X)

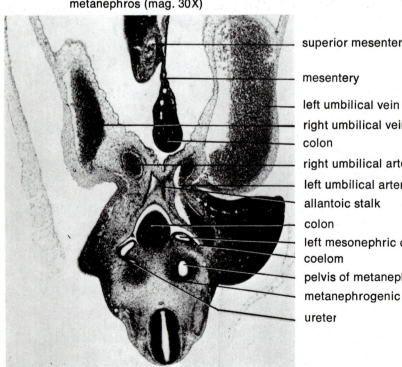

superior mesenteric artery

mesentery

left umbilical vein

right umbilical vein

colon

right umbilical artery

left umbilical artery

allantoic stalk

colon

left mesonephric duct

coelom

pelvis of metanephros

metanephrogenic mesenchyme

ureter

metanephros

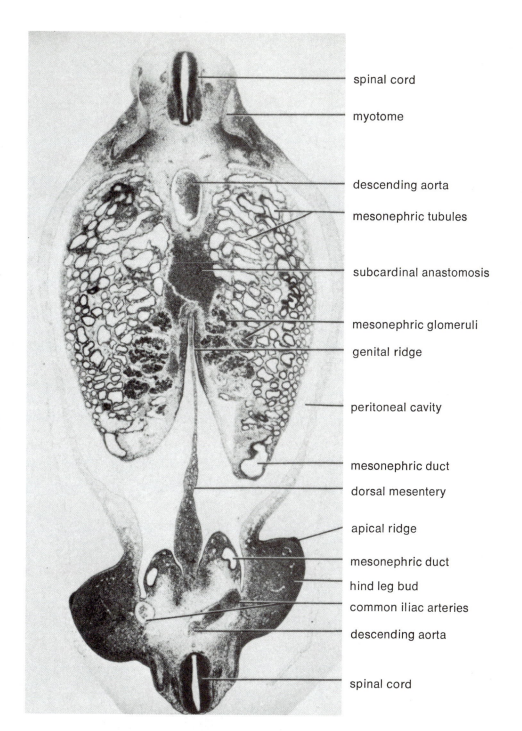

spinal cord

myotome

descending aorta

mesonephric tubules

subcardinal anastomosis

mesonephric glomeruli

genital ridge

peritoneal cavity

mesonephric duct

dorsal mesentery

apical ridge

mesonephric duct

hind leg bud

common iliac arteries

descending aorta

spinal cord

Figure 269   10-mm pig embryo, transverse section through common iliac artery (mag. 30 X)

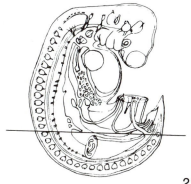

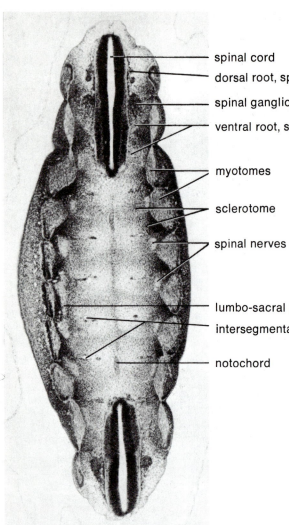

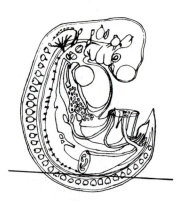

spinal cord

dorsal root, spinal n.

spinal ganglion

ventral root, spinal n.

myotomes

sclerotome

spinal nerves

lumbo-sacral plexus

intersegmental arteries

notochord

Figure 270 10-mm pig embryo, transverse section through lumbo-sacral plexus (mag. 30X)

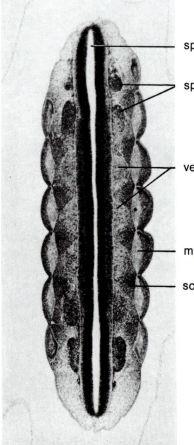

spinal cord

spinal ganglia

ventral root of spinal n.

myotome

sclerotome

Figure 271 10-mm pig embryo, transverse section through spinal nerves (mag. 30X)

20. Human Uterus and Placenta

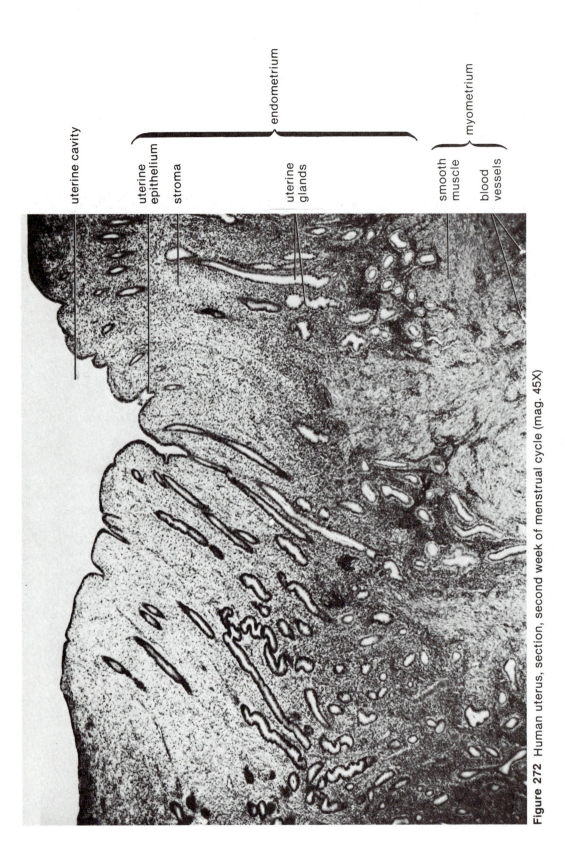

uterine cavity

uterine epithelium

stroma

endometrium

uterine glands

smooth muscle

blood vessels

myometrium

Figure 272 Human uterus, section, second week of menstrual cycle (mag. 45X)

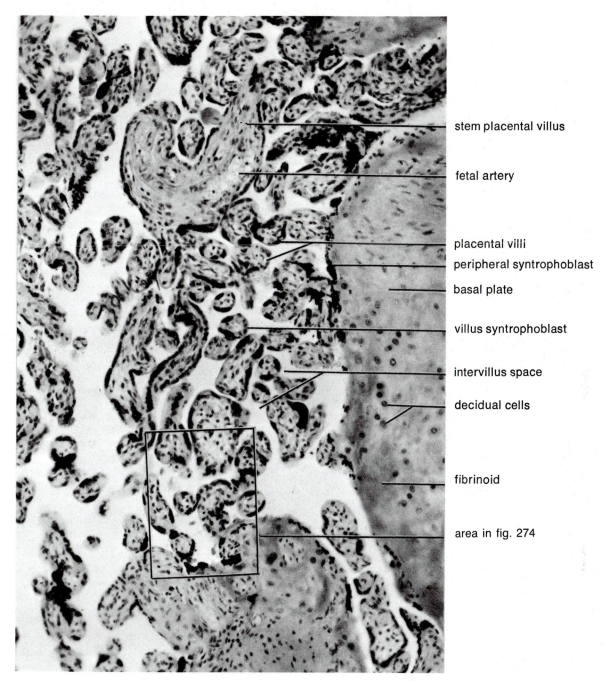

stem placental villus

fetal artery

placental villi
peripheral syntrophoblast
basal plate

villus syntrophoblast

intervillus space

decidual cells

fibrinoid

area in fig. 274

Figure 273 Human placenta, section (mag. 150 X)

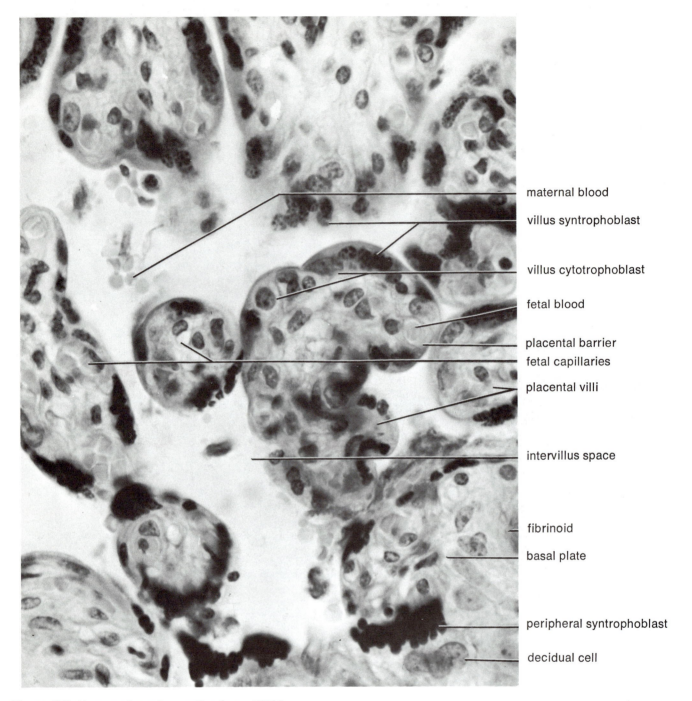

maternal blood

villus syntrophoblast

villus cytotrophoblast

fetal blood

placental barrier
fetal capillaries

placental villi

intervillus space

fibrinoid

basal plate

peripheral syntrophoblast

decidual cell

Figure 274 Human placenta, section (mag. 725X)

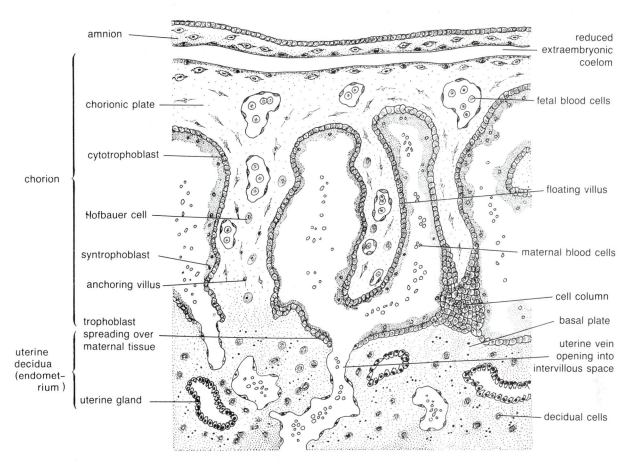

Figure 275 Semischematic drawing showing the relations of the chorionic villi and trophoblast to maternal endometrial tissues of the human placenta. (Redrawn from B. M. Patten (after Hill), *Human Embryology*. Copyright 1968 by McGraw-Hill Book Co. Used with permission of the McGraw-Hill Book Co.)

The labels in the figure are:

- amnion
- reduced extraembryonic coelom
- chorionic plate
- fetal blood cells
- cytotrophoblast
- floating villus
- Hofbauer cell
- maternal blood cells
- syntrophoblast
- anchoring villus
- cell column
- trophoblast spreading over maternal tissue
- basal plate
- uterine vein opening into intervillous space
- uterine gland
- decidual cells
- chorion
- uterine decidua (endometrium)

Glossary and Synopsis of Development

accessory cells   fig. 24

Ovarian nurse cells of sea urchins containing yolk granules, lipid droplets, and glycogen which are probably transferred to the oocytes.

accessory cleavage   fig. 147

Small cleavage furrows in the blastoderm of birds which form around supernumerary sperm nuclei; subsequently degenerate.

accessory ganglia of cranial n. 10
figs. 235, 241

A chain of small ganglia along the myelencephalon which contributes sensory fibers to the 10th and 12th cranial nerves.

acoustic ganglion   figs. 235, 239, 243
see ganglion of cranial nerve 8.

adhesive glands   figs. 97, 103, 104, 115, 123, 124

Paired ectodermal thickenings on the ventral side of the head of anuran tadpoles; secrete adhesive mucus; used for attachment.

air chamber   figs. 146, 188

An air-filled space between the inner and outer shell membranes of bird eggs; formed by shrinkage of egg contents caused by water loss through the porous shell; provides a "breathing space" for the chick before hatching.

albumen   figs. 145, 146, 188
also see chalaza

The "white" of bird eggs; provides protein and water for the growing embryo; consists of fluid and dense components including the chalaza; composed of the proteins, ovalbumin, conalbumin, ovomucin, avidin, and lysozyme; the latter two substances protect the embryo from microorganisms; produced by the magnum segment of the hen's oviduct under the stimulus of estrogen.
Syn.: egg white.

albumen sac   fig. 78
also see chorio-allantois

An extension of the chorio-allantois around the egg albumen in bird embryos; absorbs the albumen.

allantoic artery   fig. 229
also see umbilical arteries

The arterial supply of the allantois; in mammals constitutes the umbilical arteries supplying the fetal component of the placenta; arise as paired vessels connecting to the posterior ends of the dorsal aortae; cease to function at birth and degenerate.
Syn.: umbilical arteries.

allantoic cavity   fig. 188
also see allantois

The cavity of the allantois, lined by endoderm of the splanchnopleur, communicates with the cloaca via the allantoic stalk; in bird embryos and in some mammals, stores embryonic urine.

allantoic stalk   figs. 188, 228, 231, 235, 238, 262

A canal in the umbilical cord connecting the cloaca or, later, the urogenital sinus with the allantois; carries urine to the allantois in many amniotes.

allantoic vein   figs. 207, 216, 219, 225, 228
see umbilical vein

allantois   figs. 188, 207, 208, 210, 222, 224, 226, 228

An extraembryonic membrane of amniotes; grows out of the hindgut to fuse with the chorion and form the chorioallantois; vascularized by the umbilical vessels; functions as the main embryonic respiratory organ of birds, stores embryonic urine, forms the albumen sac; in mammals, contributes to the placenta.

amnio-cardiac vesicle   figs. 159, 164
also see coelom; extraembryonic coelom; pericardial coelom

Paired dilations of the coelom in the heart region (pericardial coelom) and extending laterally into the extraembryonic area where they contribute to the formation of the amnion.

amnion   figs. 187, 188, 189, 190, 200, 208, 209, 211, 220, 223, 227, 228, 237, 240, 241
also see amniotic fold

An extraembryonic membrane of amniotes which encloses the embryo and amniotic fluid; usually arises from folds of somatopleure.

amniotic cavity   figs. 187, 188, 208, 211, 220

also see amnion

The lumen of the amnion; contains amniotic fluid and the embryo.

amniotic fold   figs. 174, 175, 183, 184, 188, 200, 201, 222

also see amnion; chorion

Folds of somatopleure surrounding the embryo, arising first anterior to the head, then extending along the sides and encircling the tail; the folds are drawn over the embryo, enclosing it with two membranes, an inner amnion, and an outer chorion.

anal arms figs. 38–42
also see oral arms
Slender, paired extensions of the ventral body wall of the pluteus; supported by skeletal rods and bear ciliated bands; function to stabilize and propel the larva.

anal pit *see* proctodeum

anastomosis of basilar artery and internal carotid artery fig. 243
A ring-like linkage of basilar and internal carotid arteries encircling the infundibulum; forms the circle of Willis.

anatomoses of omphalomesenteric veins fig. 260
After the anastomosis, certain segments of the omphalomesenteric veins (vitelline veins) atrophy, leaving a single channel, the hepatic portal vein which is therefore derived from parts of both right and left omphalomesenteric veins.

anchoring villus fig. 275
also see floating villus; placental villi; stem placental villus
A placental villus attached to the uterine decidua.

animal hemisphere figs. 33, 66, 83, 84
also see animal pole; vegetal hemisphere; micromeres
That half of the egg or early embryo containing the least yolk and consequently the most active cytoplasm. The animal pole lies at its center and opposite the vegetal pole.

animal plate figs. 35, 36
A thickening of ectoderm near the animal pole of sea urchin embryos; bears long cilia of the apical tuft.

animal pole figs. 33, 34, 41, 42, 43, 49–52, 55, 65, 85
The end of the embryonic axis centered in the most active region; opposite the vegetal pole.

anterior cardinal veins figs. 117, 120, 173, 185, 186, 190, 191, 192, 196, 206, 207, 209, 211, 212, 214, 225, 226, 228, 229, 236, 238, 239, 243–246, 250, 251
The primitive, paired veins of the head and neck; drain with the posterior cardinals into the common cardinal veins; form the cerebral veins, dural sinuses, internal jugulars, and superior vena cava.
Syn.: precardinal veins.

anterior intestinal portal figs. 159, 164, 168, 170–172, 178, 183–185, 199, 203
also see midgut
In amniotes, the opening from midgut into foregut; moves posteriorly, lengthening the foregut, and meets the posterior intestinal portal at the level of the small intestine to form the yolk stalk.

anterior liver diverticulum figs. 198, 216
also see liver; posterior liver diverticulum
One of two outgrowths of the duodenum of birds which grow, branch, and anastomose to form the liver.
Syn.: cranial liver bud; dorsal liver bud.

anterior neuropore figs. 159, 163, 172
A temporary anterior opening into the neural tube; obliterated by complete closure of the prosencephalon.

anterior vena cava fig. 260
The main venous trunk vessel draining the head, neck, and forelimb regions directly into the right atrium; derived from the right common cardinal and right anterior cardinal veins. *Syn.*: superior vena cava, precava.

anterior vitelline veins fig. 186
also see vitelline veins
Paired veins draining the anterior part of the yolk sac and the sinus terminalis; flow into the right and left vitelline veins.

antrum *see* follicular cavity

anus figs. 38, 42, 54, 55, 56, 71, 73, 115, 116, 120, 130
The posterior opening of the digestive tract; derives from the blastopore in the see urchin and starfish and from the proctodeum in Ampioxus and vertebrates.

aorta *see* descending aorta

aortic arch fig. 124
also see aortic arches 1, 2, 3, 4, 5, 6
Paired arterial connections between the dorsal and ventral aortae and encircling

a

the pharynx; they lie within the branchial arches and arise from head mesenchyme; some degenerate, others persist as adult arteries.

aortic arch 1   figs. 120, 172, 173, 185, 186, 192, 193, 206, 235

The anterior member of a series of six paired arterial connections between the dorsal and ventral aortae; lie within the mandibular arch; degenerate at an early stage.

aortic arch 2   figs. 185, 186, 194, 206, 209, 210, 212, 213, 225, 228, 235

The second of a series of six paired arterial connections between the dorsal and ventral aortae; lie within the hyoid arch; degenerate at an early stage.

aortic arch 3   figs. 106, 120, 185, 186, 204, 206, 209, 210, 212–214, 225, 227, 228, 235, 236, 238, 239, 246, 248

The third of a series of six paired arterial connections between the dorsal and ventral aortae; lie within the third branchial arch; persist as part of the internal carotid arteries and, in mammals, also as part of the common carotids.

aortic arch 4   figs. 106, 206, 225, 228, 229, 235, 236, 249

The fourth of a series of six paired arterial connections between the dorsal and ventral aortae; lie within the fourth branchial arch. The right fourth aortic arch forms the arch of the aorta in birds, while the left fourth arch forms the arch of the aorta of mammals; in amphibians both right and left persist as aortae.
Syn.: systemic arch.

aortic arch 5   fig. 236

The fifth of a series of six paired arterial connections between the dorsal and ventral aortae; vestigial and inconstant; atrophies completely.

aortic arch 6   figs. 225, 235, 236, 251

The sixth of a series of paired arterial connections between the dorsal and ventral aortae; form part of the pulmonary arteries of tetrapods and, in late embryonic stages, the ductus arteriosus.

aortic sac   *see* ventral aorta

apical bronchus   figs. 255, 258

also see lateral bronchus; stem bronchus
An unpaired evagination of the trachea or the right primary bronchus which, together with the two lateral bronchi and the stem bronchi, constitute the secondary bronchi of pig and human embryos; subsequently forms the upper lobe of the right lung.

apical ridge   figs. 219, 222, 257, 267, 269
An ectodermal thickening on the margin of the limb buds of amniotes.

archenteric vesicle   fig. 52
also see coelomic sacs
The inner expanded end of the archenteron of the starfish gastrula; forms mesenchyme which scatters through the blastocoel and then divides into the coelomic sacs.

archenteron   figs. 35–37, 51–53, 67, 68, 72 73, 86–88, 90, 91
The primitive gut formed by the gastrulation movements; at least part of its wall is endoderm; opens through a blastopore to the exterior. *Syn.*: gastrocoel.

archenteron roof   fig. 87
The dorsal wall of the archenteron; in amphibians, forms the notochord and dorsal mesoderm; in Amphioxus, forms the notochord and somites.

archinephric duct   *see* mesonephric duct

area opaca   figs. 148–150, 155–158
The transluscent peripheral zone of the chick blastoderm which is attached to the yolk underneath; surrounds the area pellucida.

area opaca vasculosa   figs. 154, 159, 160
also see blood island

The inner region of the area opaca; contains thickenings of splanchnic mesoderm, the blood islands, which differentiate into red blood cells and vitelline blood vessels.

area opaca vitellina   figs. 159, 160
The outer region of the area opaca; extends outward from the area vasculosa and is temporarily free of blood and blood vessels.

area pellucida   figs. 148–150, 154, 159, 160
The relatively transparent central region of the chick blastoderm; underlaid by the fluid-filled subgerminal cavity. The primitive streak and embryonic axis form within it.

artifact   figs. 79, 85, 127, 144, 193
A change in the structure of a microscopic preparation caused by the processing technique; often appears as spaces or cracks from shrinkage or as granules and fibers precipitated by fixing fluids and solvents.

arytenoid folds   figs. 247, 249
Paired ridges lateral to the glottis; derived from the 4th and 5th branchial arches; contribute to the wall of the larynx.

ascending aorta   figs. 237, 250, 251
The part of the aorta which extends anteriorly from the heart; forms from the longitudinal division of the bulbus cordis and from the ventral aorta.

aster   fig. 28
Fibers radiating from the centrioles during mitosis; forms the mitotic apparatus together with the spindle.

atretic follicle   figs. 7, 8
A degenerating ovarian follicle in mammals. Atresia may appear at any stage of follicular growth, destroying the oocyte, follicle cells, and theca; in some species (cat) atresia involves dispersal of some follicle cells, which then form interstitial cells.

atrial septum   fig. 252
also see interatrial foramen; atrial septum I; atrial septum II
A partition arising from the wall of the primitive atrium near its sagittal plane in tetrapods; eventually divides it into right and left atria. An interatrial foramen through the septum prevents complete separation of the atria until its closure after hatching in birds or after birth in mammals.

atrial septum I   fig. 254
also see atrial septum II, interatrial foramen
A creseent-shaped median partition of the atrium dividing it into a right and left atrium; the atrial septum I is incomplete; the original opening through it is the

interatrial foramen I which allows blood to pass from the right to left atrium; this foramen closes early but at the same time is replaced by the interatrial foramen II; contributes to the adult atrial septum.

atrial septum II   fig. 254
also see atrial septum I; interatrial foramen
A second atrial septum which forms later

and on the right side of the atrial septum I; it overlaps the interatrial foramen II but is also perforated by the foramen ovale; the atrial septa I and II fuse to form the adult atrial septum. The foramen ovale and interatrial foramen II do not close until after birth.

atriopore   fig. 56
also see atrium
In Amphioxus, a ventral opening of the atrium in the posterior one third of the body; water entering the oral hood passes into the pharynx, through the gill slits into the atrium and out the atriopore.

atrioventricular canal   figs. 253, 254, 255
The passage within the heart connecting the atrium with the ventricle; encircled and constricted by the endocardial cushion; divides during the partitioning of the heart.

atrium   figs. 56, 107, 119, 124, 135, 170, 172, 173, 183-187, 196, 203, 204, 206-208, 210, 215, 224, 227, 228, 232, 234
also see left atrium; right atrium

The heart chamber which in the embryo receives blood from the sinus venosus and delivers it to the ventricle; in tetrapods, becomes partitioned into right and left atria and incorporates the sinus venosus; in Amphioxus, a chamber around the pharynx.

auditory nerve   *see* ganglion of cranial nerve 8

auditory placode   *see* otic placode

auditory vesicle   *see* otic vesicle

axial filament   fig. 15
also see centrioles; sperm tails
A central fiber in cilia and the tail or flagellum of sperm; composed of a ring of nine double microtubules and a central pair of microtubules; appears to arise from the centrioles near the head of the sperm; functions in the tail movements.

basal plate   figs. 273-275
also see peripheral syntrophoblast; decidual cells; fibrinoid
A sheet of chorionic tissue in contact with the maternal decidua of the placenta; composed of peripheral syntrophoblast, peripheral cytotrophoblast, fibrinoid; forms part of the afterbirth.

basement membrane figs. 2, 5, 6, 142
A fibrous sheet beneath many kinds of epithelia; supports the epithelia.

basilar artery figs. 225, 235, 236, 242-245
A median vessel beneath the hindbrain; connects the vertebral arteries with the internal carotids.

belly stalk fig. 188
The slender link between bird embryos and their extraembryonic membranes; covered by amnion and contains the vitelline and allantoic blood vessels; homologous to the umbilical cord of mammals.

bivalent fig. 17
A pair of homologous chromosomes in synapsis; found within the primary spermatocyte or primary oocyte during the 1st maturation division. *Syn.*: tetrad.

blastocoel figs. 33-37, 48-52, 54-57, 58, 61, 65, 66, 72, 74, 126
The cavity of the blastula; a closed space which is invaded early by mesenchyme and in some species is greatly reduced or obliterated by the enlarging archenteron. *Syn.*: segmentation cavity.

blastoderm figs. 145, 146
That stage in the development of telolecithal eggs (some fish, reptiles, birds) in which the embryo consists of a nearly flat sheet of cells spreading over the surface of the yolk; corresponds to the blastula stage of homolecithal eggs.

blastopore figs. 36, 51-53, 67, 74, 86, 87, 89, 90, 91
The opening of the archenteron to the exterior; marks the point of origin of the archenteron and the caudal end of the embryo; forms anus in echinoderms.

blastula figs. 33, 34, 84, 85
The final stage of cleavage—typically, a hollow sphere—with the embryonic cells surrounding a cavity, the blastocoel.

blood island figs. 154, 159, 182
A mass of splanchnic mesodermal cells in the gut wall of amphibians or the yolk sac of amniotes; the first blood-forming tissue of the embryo, forms red blood cells and vitelline blood vessels.

blood vessels figs. 1, 2, 7, 8, 272
also see medulla; myometrium

Blood vessels extend into organs and spread through them following strands or sheets of connective tissue such as those of the testis, medulla of the ovary and of the myometrium.

body wall figs. 183, 184, 187, 208, 216, 228, 251, 256
The outer layer of the body, enclosing the body cavities and viscera; derives from epidermis, dermatome, myotome, and somatic mesoderm of the embryo.

brachial plexus figs. 234, 259, 261
An interconnection of ventral rami of cervical and thoracic spinal nerves from which branch the nerves of the forelimb.

brain vesicle figs. 69, 70, 73
also see neural tube
In Amphioxus, the slightly enlarged anterior end of the neural tube.

branchial arch 1 *see* mandibular arch

branchial arch 2 *see* hyoid arch

branchial arch 3 figs. 228, 247
also see aortic arch 3
The third of a series of paired bars in the wall of the pharynx, within which are formed the third aortic arch and a cartilage bar of the visceral skeleton; in aquatic vertebrates, forms and supports gills; in mammals, contributes to the epiglottis and in tetrapods to the hyoid bone.

branchial arch 4 figs. 118, 213, 247
also see aortic arch 4
The fourth of a series of paired bars in the wall of the pharynx, within which are formed the fourth aortic arch and a cartilage bar of the visceral skeleton; in aquatic vertebrates, forms and supports gills; in tetrapods, contributes to the larynx, and in mammals, to the epiglottis.

branchial cleft 1 figs. 172, 183, 184, 193, 203, 204, 207, 209, 210, 224, 230, 233, 234
also see branchial groove 1; pharyngeal pouch 1
A slit-like perforation in the wall of the pharynx between the mandibular and hyoid arches; formed from first pharyngeal pouch and first branchial groove; forms the eustachian tube, middle ear cavity, and external ear canal; in lower vertebrates (frog), contributes to the thymus.

Syn.: gill slit 1; hyomandibular cleft; visceral cleft 1.

branchial cleft 2   figs. 103, 183, 184, 194, 203, 204, 207, 209, 210, 224, 230
also see branchial groove 2; pharyngeal pouch 2
A slit-like perforation in the wall of the pharynx between the hyoid and third branchial arches; formed from the second pharyngeal pouch and second branchial groove; subsequently closes and obliterates, except contributes to the thymus in lower vertebrates (frog).
Syn.: gill slit 2; visceral cleft 2.

branchial cleft 3   figs. 103, 203, 207, 209, 210, 224, 230, 239
also see pharyngeal pouch 3
A slit-like perforation in the wall of the pharynx between the third and fourth branchial arches; formed from the third pharyngeal pouch and third branchial groove; subsequently closes, parts of its wall contributing to the thymus and parathyroids. *Syn.*: gill slit 3; visceral cleft 3.

branchial cleft 4   figs. 207, 210, 224
also see branchial groove 4; pharyngeal pouch 4
A thin region in the wall of the pharynx between the fourth and fifth branchial arches; formed from the fourth pharyngeal pouch and fourth branchial groove; breaks through only in lower vertebrates (frog); contributes to the parathyroids.
Syn.: gill slit 4; visceral cleft 4.

branchial groove 1   figs. 192, 213, 214, 244–246
also see branchial cleft 1
An ectodermal invagination meeting the first pharyngeal pouch to form the first branchial cleft; forms the external ear canal in amniotes. *Syn.*: visceral groove 1.

branchial groove 2   figs. 214, 249
also see branchial cleft 2
An ectodermal invagination meeting the second pharyngeal pouch to form the second branchial cleft; obliterates in tetrapods. *Syn.*: visceral groove 2.

branchial groove 4   figs. 201, 248
also see branchial cleft 4
An ectodermal invagination meeting the fourth pharyngeal pouch to form the fourth

branchial cleft; obliterates in tetrapods.
Syn.: visceral groove 4.

branchial pouch   *see* pharyngeal pouch

bulbus cordis   figs. 107, 117, 120, 124, 170, 172, 173, 176, 183–187, 195, 196, 203, 204, 206–209, 215, 224, 226, 227, 232, 234, 237, 252
The heart chamber, originally most anterior in position, connecting the ventricle with the ventral aorta; in tetrapods, partitioned longitudinally to form the ascending aorta and pulmonary aorta. *Syn.*: bulbus arteriosus; conus arteriosus.

caecum   figs. 199, 266
A pocket in the large intestine of mammals near its connection to the small intestine; appears as an enlargement on the intestinal loop early in the development of the intestine

cardiac jelly   fig. 169
A gelatinous substance between the epicardium and epimyocardium of the early embryonic heart; maintains the separation between the heart layers for a period of time; eventually replaced by proliferating cells of heart wall.

caudal artery   figs. 120, 235, 236
The extension of the aorta into the tail.

caudal fin   figs. 56, 71
also see dorsal fin
The tail fin.

caudal flexure   figs. 207, 224
The ventral bend in the tail and posterior trunk; together with the flexures in the head and anterior trunk, forms the embryo into a compact C-configuration.

caudal liver bud   *see* posterior liver diverticulum

caudal vein   fig. 120
The principal vein of the tail; drains into the posterior cardinals.

cell center   *see* centrosome

centrioles   figs. 11, 15
also see sperm tails; axial filament
A pair of minute cytoplasmic granules, usually near the nucleus, surrounded by a zone of gelated cytoplasm; self-replicating; composed of a ring of nine sets of ultramicroscopic tubules; associated

C

with the formation of spindle fibers and axial filaments of cilia and flagellae.

centrosome   figs. 18, 19
A cytoplasmic organelle composed of two minute granules, the centrioles, and surrounding fibrous protein; forms the mitotic spindle in dividing cells and in sperm also the axial filament. *Syn.*: cell center.

cerebral aqueduct   figs. 187, 194, 226, 237
also see fourth ventricle; third ventricle
The neural canal of the mesencephalon; the originally large chamber is progressively narrowed by the thickening of the walls of the midbrain, becoming a slender canal connecting the third and fourth ventricles.

cerebral hemispheres   figs. 210, 216, 224, 228, 234, 251
also see telencephalon
Paired dorso-lateral bulges of the telencephalon; form the cerebrum.

cervical flexure   fig. 207
also see caudal flexure; cranial flexure; dorsal flexure
One of several ventral bends in the body axis giving the amniote embryo a compact C-configuration; forms in region of the hindbrain and anterior trunk.

cervical segmental artery   figs. 225, 236
One of a series of small branches of the aorta arising between the cervical somites; contributes to the subclavian and vertebral arteries.

cervical sinus   figs. 204, 233, 248
A depressed region in the sides of the neck bearing the third and fourth branchial grooves in its floor; subsequently closes and obliterates.

chalaza   fig. 146
also see albumen
Strands of dense albumen of the bird egg attached to the vitelline membrane of the yolk holding it near the center of the egg yet allowing it to rotate and float upward to a position near the upper shell surface favorable for incubation.

chitinous layer   figs. 17–21
A thick, clear layer of the egg shell of *Ascaris;* formed from the egg after separation of the fertilization membrane; composed of chitin and protein.

chorda   *see* notochord

chordamesoderm   figs. 60, 87, 88, 90
The archenteron roof of amphibian gastrulae; arises by involution of marginal zone cells over the lips of the blastopore; forms the notochord in the dorsal position and the mesoderm laterally and ventrally.

chorda-neural crescent   figs. 72, 74
also see mesodermal crescent
In Amphioxus embryos, during cleavage a zone on the dorsal side of the embryo opposite the mesodermal crescent; the part lying above the equator will subsequently form the nervous system, the part of the crescent below the equator will form notochord after gastrulation.

chorio-allantois   figs. 188, 229
A composite structure formed by the fusion of the chorion and allantois; the principle respiratory organ of bird embryos; forms the albumen sac of bird embryos and absorbs the albumen; forms the embryonic component of the placenta of most mammals; vascularized by the allantoic blood vessels.

chorion   figs. 188–190, 192, 195–197, 199, 200, 209, 211, 215, 218, 220–223, 229, 275
also see amniotic fold
The outermost extraembryonic membrane of amniotes; arises from somatopleure and is usually drawn over the embryo by the amniotic folds; later fuses with the allantois to form the chorioallantois, which is vascularized by the umbilical vessels and functions as the main embryonic respiratory organ of birds; the chorio-allantois contributes to the placenta of mammals. *Syn.*: serosa.

chorionic plate   fig. 275
also see placental villi
The basal layer of the chorionic component of the placenta from which chorionic villi arise.

choroid fissure   figs. 185, 194, 207, 213
A groove in the ventral wall of the optic cup and optic stalk; after mesenchyme and blood vessels invade the fissure, its lips fuse.

chromosomes   figs. 18, 79
Threads of chromatin in the cell nucleus or in the mitotic and meiotic division figures; contain the genes in linear order; composed of DNA, RNA, and protein.

cicatrix   fig. 145
A thin avascular strip in the follicular wall of bird ovaries; rupture of the mature follicle to release the egg during ovulation occurs through the cicatrix.

ciliated band   figs. 39, 41, 42, 54, 55
Continuous strips of tall ciliated cells in the epidermis of echinoderm larvae (pluteus, bipinnaria) extending over the body and arms; function in locomotion and food gathering.

cleavage division, 1st   fig. 20
The first mitotic division of the egg after fertilization; forms the two-cell stage.

cleavage division, 2nd   fig. 21
The second mitotic division after fertilization; forms the four-cell stage.

cleavage furrow   figs. 83, 84
A constriction in the cytoplasm which divides the egg or blastomere; forms during the telophase of the cleavage divisions; the constriction is formed by a ring of actin microfilaments.

cleavage stage   figs. 20, 21, 29–34, 83–85
The period of development beginning with the first mitotic division of the egg and ending with the blastula; a period of rapid mitoses during which no growth occurs, the cells becoming smaller, as they become more numerous.

cloaca   figs. 120, 128, 130, 140, 145, 203, 209, 224, 228, 263, 266
also see urogenital sinus
The posterior chamber of the vertebrate digestive tract; receives the allantoic stalk, urinary ducts, and reproductive ducts; partitioned in mammals to form the rectum, urinary bladder, and urogenital sinus.

cloacal membrane   figs. 223, 238, 263
A double-layered membrane formed where the ventral wall of the cloaca fuses with ventral ectoderm; ruptures to open the anus and, in mammals, the urogenital sinus as well.

club-shaped gland   figs. 70, 71, 73
also see club-shaped gland duct
In Amphioxus, a temporary larval organ arising as an evagination of the gut between the endostyle and pharynx; forms an external opening ventral to the mouth;

degenerates at the larval stage of 12 gill slits.

club-shaped gland duct   fig. 73
also see club-shaped gland
The duct is located just ventral to the mouth.

coeliac artery   figs. 225, 235, 236
A ventral branch of the aorta supplying the anterior digestive system; derives from vitelline arteries.

coelom   figs. 24, 99, 113, 164, 165, 173, 176, 180, 181, 182, 202, 208–210
also see coelomic vesicles; embryonic coelom; extraembryonic coelom; pericardial coelom
A cavity within the mesoderm which forms the body cavities of the adult; in vertebrates, arises as a cleft in the lateral plate mesoderm, which is thereby divided into somatic and splanchnic layers.

coelomic sac   figs. 38, 41, 42, 53–55
also see archenteric vesicle
Thin-walled, mesodermal evagination of the archenteron of echinoderm gastrulae (sea urchin, starfish); forms the coelom and water vascular cavities of the adults.

collecting tubules   fig. 77
In the frog testis, conducting vessels carrying mature sperm from the seminiferous tubules to the vasa efferentia.

colon   figs. 235, 267, 268
also see caecum
Part of the large intestine; arises from hindgut.

common bile duct   figs. 262, 264
A vessel connecting the cystic and hepatic ducts to the duodenum; conducts bile into the duodenum; arises from the stem of the liver diverticulum. *Syn.*: ductus choledochus.

common cardinal vein   figs. 120, 173, 185, 197, 206, 207, 209, 215, 224, 226, 229, 236, 238, 239, 252, 253, 255, 260
The trunk of the anterior and posterior cardinal veins; connects to the sinus venosus; contributes to the anterior vena cava of the adult, and the oblique vein. *Syn.*: duct of Cuvier.

common iliac arteries   figs. 225, 236, 269
also see umbilical arteries
The large terminal branches of the aorta; arise from the proximal segments of the

C

umbilical arteries and a pair of dorsal
intersegmental arteries; persist as the
common trunks of the external and
internal iliac arteries.

connective tissue figs. 8, 141, 144
One of the main kinds of tissue;
characterized by much intercellular material,
including fibers. *Syn.*: stroma.

conus arteriosus *see* bulbus cordis

copula figs. 246, 247
also see tongue
A median elevation on the floor of the
mouth arising from the hyoid arch and
contributing to the root of the tongue.

coronary sinus fig. 260
A venous trunk vessel within the dorsal wall
of the heart which drains the coronary
veins of the heart wall into the right atrium;
derives from a persistant left common
cardinal vein.

corpus luteum *(pl.* **corpora lutea***)* fig. 9
A mass of endocrine gland tissue in the
ovary of viviparous vertebrates; forms from
an ovulated follicle and persists into
pregnancy; secretes progesterone.

cortex figs. 7, 9, 58
The outer zone of an organ or egg;
contains the follicles and corpora lutea
of the ovary.

cranial cartilage fig. 134
Cartilage bars forming part of the primitive
skull or chondocranium; arise from
mesodermal head mesenchyme and neural
crest; eventually replaced by bony skull in
higher vertebrates.

cranial flexure fig. 207
also see caudal flexure; cervical flexure;
dorsal flexure
One of several ventral bends in the body
axis giving the amniote embryo a compact
C-configuration; forms in the midbrain;
the only permanent flexure.

cranial nerve 3 figs. 213, 214, 229, 235,
241–243
A pair of motor nerves arising from the
floor of the mesencephalon; innervate
some extrinsic and all inner eye muscles.
Syn.: oculomotor nerve.

cranial nerve 4 figs. 235, 240
A pair of motor nerves arising from the

mesencephalon and emerging from the
roof of the brain at the isthmus; innervate
the superior oblique ocular muscles.
Syn.: troclear nerve.

cranial nerve 5 figs. 212, 242
also see ganglion of cranial nerve 5;
mandibular ramus of cranial nerve 5;
maxillary ramus of cranial nerve 5;
ophthalmic ramus of cranial nerve 5;
root of cranial nerve 5
A pair of mixed nerves arising from the
sides of the metencephalon and semilunar
ganglia; three divisions—the ophthalmic,
maxillary, and mandibular rami—innervate
the mandibular arch region.
Syn.: trigeminal nerve.

cranial nerve 6 fig. 244
A pair of motor nerves emerging from
the floor of the myelencephalon; innervate
the external rectus eye muscles.
Syn.: abducens nerve.

cranial nerve 7 figs. 212, 235,
242–245
also see ganglion of cranial nerve 7
A pair of mixed nerves arising from the
myelencephalon at the anterior margin
of the otic vesicle and the geniculate
ganglion; innervate the hyoid arch.
Syn.: facial nerve.

cranial nerve 9 figs. 243, 244
also see ganglion of cranial nerve 9;
superior ganglion; petrosal ganglion
A pair of mixed nerves arising from the
myelencephalon at the caudal margin of
the otic vesicles; they bear the superior
and petrosal ganglia and innervate the
third branchial arch.
Syn.: glossopharyngeal nerve.

cranial nerve 10 figs. 235, 243–246, 248
also see ganglion of cranial nerve 10;
jugular ganglion; accessory ganglia of
cranial nerve 10; nodose ganglion
A pair of mixed nerves arising from the
myelencephalon and bearing the jugular
and nodose ganglia; innervate branchial
arches 4, 5, and 6, and extend
parasympathetic fibers to the viscera;
in aquatic vertebrates (frog tadpole),
innervate the lateral line.
Syn.: vagus nerve.

cranial nerve 11 figs. 242–244
A pair of motor nerves arising from the

236

myelencephalon and spinal cord and innervating muscles of the pharynx and shoulder; part of the vagus nerve in aquatic vertebrates. *Syn.*: accessory nerve.

cranial nerve 12   figs. 235, 237, 243–246, 248

A pair of nerves arising from many rootlets on the ventral wall of the myelencephalon; innervate the tongue muscles; evolved from cervical spinal nerves of aquatic vertebrates. *Syn.*: hypoglossal nerve.

cumulus oophorus   figs. 7, 8
also see stratum granulosa
A thickening in the stratum granulosa containing the mammalian oocyte.

cystic duct   fig. 264
also see gall bladder
The duct of the gall bladder connecting it with the common bile duct; arises from the liver diverticulum.

cytoplasm   figs. 27, 43
The part of the cell or egg (oocyte) excluding the nucleus; contains more or less yolk in eggs.

decidual cells   figs. 273–275
Enlarged connective tissue cells of the maternal decidua; distributed beneath the basal plate; contain abundant stores of glycogen and lipid.

decidua of uterus   fig. 275
also see endometrium; placental villi
The inner superficial zone of the endometrium; during pregnancy is in direct contact with the placental villi; detached during the after contractions following birth and is expelled as part of the afterbirth.

dense mesenchyme   figs. 36, 246,
A condensation of mesenchyme cells preparatory to the formation of cartilage in these locations.

dermatome   figs. 126, 175, 199, 215
also see somites
The outermost division of the somite; lies under and in contact with epidermis forming the dermis or connective tissue of the skin.

descending aorta   figs. 186, 187, 196, 197, 203, 204, 206, 208, 212, 213, 219, 224, 226, 227, 232, 236, 237, 259, 265, 266, 269

also see dorsal aortae
The principal artery of the trunk; a median vessel formed by the fusion of the paired dorsal aortae; extends from the subclavian to the common iliac arteries. *Syn.*: aorta.

diencephalon   figs. 121, 122, 131, 133, 183, 184, 187, 190–194, 203, 204, 207, 210, 215, 216, 224, 227, 230, 232, 234, 239, 244, 245, 249

The posterior division of the prosencephalon; a deep, laterally compressed region to which the optic stalks, infundibulum, and epiphysis attach; its cavity is the third ventricle of the brain; it forms the epithalamus, thalamus, and hypothalamus; its roof forms the choroid plexus.

differentiating spermatid   figs. 2, 10, 15
also see spermatid
The spermatid during its transformation into a sperm, a process called spermiogenesis, during differentiation the spermatid is embedded in a pocket within the Sertoli cell. *Syn.*: immature sperm.

diffuse stage   fig. 143
A phase of the diplotene stage of meiosis in some species; the chromosomes become extended and diffuse resembling the interphase state.

diplotene stage   figs. 3, 14, 143
also see pachytene stage
A stage of the first maturation division in spermatogenesis and oogenesis; follows the pachytene stage; chromosomes in synapsis separate except at points of crossing over (chiasmata); followed by diakinesis during which the chromosomes separate farther moving the chiasmata to the ends of chromosomes (terminalization); metaphase of the first maturation division follows diakinesis.

dorsal aorta   figs. 120, 126–128, 129, 131, 138, 140, 170–173, 175, 176, 179, 185, 192, 193, 195, 200, 206, 207, 210, 212, 220, 221, 227, 228, 235, 245–248, 250, 255, 257
also see descending aorta
The primitive paired, longitudinal arteries of the trunk which fuse together posterior to the pharynx to form the descending aorta; in pharyngeal region, contribute to the internal carotids, descending aorta, and, in mammals, the right subclavian artery.

dorsal diverticulum   figs. 67, 73
also see left diverticulum; right diverticulum
In Amphioxus, a dorsal evagination of the anterior end of the gut; extends ventrally dividing into left and right diverticula which separate from the gut.

dorsal fin   figs. 56, 114, 128, 129
A flat extension of the body wall along the dorsal midline of the trunk and tail; degenerates during metamorphosis in anurans.

dorsal flexure   fig. 207
also see caudal flexure; cervical flexure; cranial flexure
One of several ventral bends in the body axis giving the amniote embryo a compact C-configuration; forms in the trunk region.

dorsal lip   figs. 53–56, 67, 74, 76–77, 79
also see ventral lip; lateral lip
The margin of the blastopore toward the animal pole and at the dorsal side of the embryo; the first blastopore lip to form in amphibians, derives from the gray crescent area of amphibian eggs; forms the foregut roof, notochord, and dorsal mesoderm; in Amphioxus forms notochord.

dorsal liver bud   *see* anterior liver diverticulum

dorsal mesentery   figs. 198, 199, 216–218, 265, 266, 269
also see greater omentum
A double layer of splanchnic mesoderm suspending the gut from the dorsal body wall and extending from the esophagus to the cloaca; provides a path and support for nerves and vessels of the gut; forms the mediastinum, greater omentum, and mesenteries of the intestine; contributes to the diaphragm.

dorsal mesocardium   figs. 169, 177, 195, 197, 215
The dorsal mesentery of the heart; derives from the ventral mesentery of the foregut; soon ruptures and disappears.

dorsal mesoderm   figs. 94, 169
also see head mesenchyme; prechordal plate; segmental mesoderm; somites
The thickened axial and paraxial mesoderm; closely associated with the neural plate and neural tube; extends from the prechordal plate and head mesenchyme

through the somites and segmental mesoderm of the trunk.

dorsal pancreas   figs. 217, 235, 263–265
also see ventral pancreas
A dorsal evagination of the duodenum; together with a ventral evagination (two in frogs and birds), forms the rudiments of the pancreas, which fuse to form one glandular mass.

dorsal root ganglia   *see* spinal ganglia

dorsal root of spinal nerve   figs. 259, 260, 270
also see spinal ganglia
The dorsal division of a spinal nerve connecting the trunk of the nerve to the alar plate of the spinal cord; composed of sensory nerve fibers; bears the dorsal root ganglion. *Syn.*: sensory root.

duct of Cuvier   *see* common cardinal vein

ductus choledochus   *see* common bile duct

ductus venosus   figs. 207, 208, 216, 226, 227, 236, 237, 239, 259, 260
A vein within the liver of amniotes carrying blood from the vitelline and left umbilical veins to the sinus venosus; derives from vitelline veins; obliterates after hatching in birds or after birth in mammals.

duodenum   figs. 217, 263, 264
The first segment of the small intestine; arises from foregut; forms the diverticula of the liver and pancreas; later connects with the common bile duct and pancreatic ducts.

ear placode   *see* otic placode

ear vesicle   *see* otic vesicle

ectoderm   figs. 67, 75, 88, 90, 162–164, 167, 169, 173, 179, 181, 182, 185, 188
also see head ectoderm; nervous layer of ectoderm
The outermost of the three primary germ layers; develops into epidermis, skin glands, hair, feathers, nails, scales, nervous system, lining of the nose, inner ear, retina and lens of the eye, pituitary gland, mouth, pigment cells, anus.

egg pronucleus   fig. 24
also see mature ova
The haploid nucleus of the mature egg formed by the completion of the second

maturation division.

egg shell   fig. 20
also see shell
The outer protective covering of eggs; in *Ascaris* the shell consists of an outer uterine layer derived from uterine fluid, a fertilization membrane and an inner chitinous layer from the egg.

embryonic coelom   figs. 178, 179, 183, 184, 187, 197, 198, 200, 201, 219, 222, 226, 228
also see coelom; extraembryonic coelom
The division of the coelom within the head and body folds; forms the pericardial, pleural, and peritoneal cavities of the adult.

embryonic shield   fig. 148
A thickening of the blastoderm during the pre-streak stage of chick embryos; marks the posterior end of the future embryonic axis.

endocardial cushion   figs. 253, 255
A ring of connective tissue which encircles and then divides the atrioventricular canal; forms the atrioventricular valves.

endocardial tube   fig. 168
also see endocardium; heart
The thin tubular lining of the embryonic heart; fuses with the epimyocardium later to form the heart wall.

endocardium   figs. 168, 169, 177, 196, 197, 215, 216
The lining of the heart; arises from splanchnic mesoderm; fuses with the epimyocardium to form the heart wall.

endoderm   figs. 67, 68, 88, 90, 91, 93–95, 98, 162–164, 166, 167, 181, 182, 188, 219
also see hypoblast; yolk endoderm
The innermost of the three primary germ layers, inward movement of which is part of gastrulation; forms the lining of the digestive and respiratory tracts, the pancreas, liver, thyroid, parathyroids, thymus and primordial germ cells (except in urodeles), the bladder, and urethra; and in Amphioxus, the head cavity, preoral cavity, endostyle, club-shaped gland.
Syn.: entoderm.

endolymphatic duct   fig. 241
The stalk of the otic vesicle; except in elasmobranchs, soon loses its connection with the body surface; forms part of the inner ear.

endometrium   fig. 272
also see uterine epithelium; uterine glands
The glandular membrane lining the uterus; consists of uterine epithelium, glands and connective tissue stroma with blood vessels; shows a marked sequence of changes during the menstrual cycle (growth, secretion, degeneration or menstruation, repair) which is regulated by ovarian hormones; receives and nourishes the embryo during pregnancy.

endostyle   figs. 70, 71, 73
In Amphioxus, a ciliated groove in the floor of the pharynx; secretes and transports mucous with food toward the esophagus; arises as a thickening in the right anterior wall of the gut.

enterocoel   fig. 75
also see somite
The cavity of the somites as in Amphioxus; derived from the archenteron; forms the myocoel and ventrally the entrocoels of somites fuse to form coelom.

ependymal layer   figs. 241, 259, 261
The inner layer of primitive neuroepithelial cells of the neural tube; by proliferation of cells, supplies neuroblasts for the mantle and marginal layers and subsequently forms the ependyma of the spinal cord and brain.

epiblast   figs. 149, 150, 153, 158
The upper or outer layer of the blastoderm of birds and mammals; in the area pellucida of birds the original caudal half of the epiblast migrates through the primitive streak to form endoderm and mesoderm; the anterior half of the epiblast becomes ectoderm.

epibranchial placodes   fig. 112
Ectodermal thickenings dorsal to the branchial clefts; contribute cells to the cranial ganglia.

epidermal overgrowth   fig. 74
In Amphioxus, the spread of epidermal cells over the neural groove and blastopore, forms the neurenteric canal and neuropore.

epidermis   figs. 91, 94, 95, 98, 99, 108
The outer, epithelial layer of the skin; derives from the ectoderm.

epiglottis   figs. 246, 247

An elevation on the floor of the pharynx anterior to the glottis in mammals; composed of cartilage in the adult; derives from branchial arches 3 and 4.

epimere   *see* somites

epimyocardium   figs. 168, 169, 172, 177, 196, 197, 215, 216

The outer layer of the heart, including the heart muscle; forms from splanchnic mesoderm; fuses with the endocardium to form the heart wall.

epiphysis   figs. 103–105, 115, 120, 121, 185, 203, 204, 207, 208, 224

An evagination from the roof of the diencephalon; forms the pineal gland.

esophageal plug   fig. 125

A mass of cells temporarily blocking the esophagus before the amphibian larva begins to feed.

esophagus   figs. 38–41, 54, 55, 131, 137, 138, 215, 226, 237, 252–257

Part of the digestive tract that connects the pharynx (or mouth in the sea urchin) with the stomach; lengthens markedly in amniotes during development; arises from foregut in vertebrates and from archenteron in the sea urchin; forms the crop of birds.

exocoel   *see* extraembryonic coelom

external carotid artery   figs. 120, 225, 236, 248

An artery of the head; arises as an outgrowth of the ventral aorta near the base of aortic arch 3; initially, supplies the mandibular and hyoid arches.

external gill   fig. 118

also see gill

Respiratory organs growing out of branchial arches 3 to 6 in amphibia aerating blood from the aortic arches; later covered by the operculum and replaced by internal gills.

external jugular vein   figs. 120, 249

A vein of the head; arises as a branch of the anterior cardinal; initially, drains the mandibular and hyoid arches.

external layer   figs. 20, 21

The outermost layer of the egg shell of *Ascaris;* adherent to the fertilization membrane; formed from uterine secretion.

external nares   fig. 130

also see internal nares; olfactory pit

The external opening of the nasal passage; derived from the olfactory pit.

extraembryonic coelom   figs. 174, 175, 178, 179, 188, 190, 200, 201, 211, 215, 220, 222, 275

also see coelom

The division of the coelom outside the head and body folds; lies between the chorion and amnion and between chorion and yolk sac; in chick and pig, receives the expanding allantois. *Syn.*: exocoel.

eye cup   *see* optic cup

eye vesicle   *see* optic vesicle

facial nerve   *see* cranial nerve 7

falciform ligament   figs. 239, 262

The ventral ligament of the liver; attaches the liver to the ventral body wall; derives from the ventral mesentery.

fertilization membrane   figs. 17–19, 28–31, 44–47, 57, 86, 87

A membrane separated from the surface of the egg after fertilization in many aquatic species; derives from the vitelline membrane, often thickened by material from the cortical granules; contributes to the egg shell in *Ascaris.*

fetal artery   fig. 273

also see stem placental villus

In the placenta, a branch of the umbilical artery found in the placental plate and in the villi; supplies the capillaries of the villi with fetal blood.

fetal blood   fig. 274

also see fetal capillaries; fetal artery

In the human placenta, the blood within the umbilical and placental vessels; separated from maternal blood by the placental barrier except during labor when tears in the barrier may permit some mixing.

fetal capillaries   fig. 274

also see placental barrier

In the placenta, the anastomosing capillary bed of the placental villi; contains fetal blood supplied by the umbilical arteries and drained by the umbilical vein; forms part of the placental barrier.

fibrinoid   figs. 273, 274

also see basal plate

An intercellular matrix of the basal plate in which peripheral cytotrophoblast cells are frequently embedded; gives a positive periodic acid-Schiff reaction for polysaccharide.

floating villus   fig. 275
also see placental villi
Free placental villi attached only at their base to the chorionic plate; suspended in the maternal blood of the intervillus spaces.

follicle cells   figs. 7, 79
also see stratum granulosa
The epithelial cells enclosing the oocyte; probably regulate the transfer of materials to the oocyte; form the stratum granulosa of bird and mammalian follicles.

follicular cavity   figs. 7–9
The space within the Graafian follicle; filled with a viscous follicular fluid.
Syn.: antrum.

forebrain   *see* prosencephalon

foregut   figs. 93, 94, 112, 118, 126, 159, 161, 163, 164, 169–172, 198, 199
also see pharynx
The part of the gut extending into the head from midgut; in amniotes, formed by the head fold as it passes under the head and trunk; eventually forms the pharynx, respiratory tract excepting the nasal passages, esophagus, stomach, duodenum, liver, and pancreas.

foreleg bud   figs. 231, 232–234, 253, 256, 257, 259, 261
also see hindleg bud; leg bud
The rudiment of the foreleg; arises as a thickening of somatic mesoderm in the body wall, an ectodermal thickening, the apical ridge forms on its margin; homologous to the wing bud of birds and arm bud of humans.

fourth ventricle   figs. 187, 208, 226, 237
also see cerebral aquaduct
The enlarged neural canal of the rhombencephalon; connected anteriorly to the cerebral aquaduct and posteriorly to the central canal of the spinal cord; its thin roof forms the choroid plexus of the medulla.

Froriep's ganglion   figs. 235, 244
In some mammals, the most caudal of the accessory cranial ganglia; may contribute to the 12th cranial nerve.

gall bladder   figs. 121, 235, 237, 263, 264
A sac-like vessel connected by the cystic duct to the common bile duct; arises from a caudal extension of the liver diverticulum; stores bile.

ganglion of cranial nerve 5   figs. 117, 123, 135, 184, 190, 191, 204, 207, 209–213, 224, 227, 230, 231, 234
also see cranial n. 5; semilunar ganglion
Ganglion arising from anterior neural crest and from the epibranchial placode above the first branchial cleft, also from the dorsolateral placodes in lower vertebrates; supplies sensory fibers to the 5th cranial nerve. *Syn.*: Gasserian ganglion; semilunar ganglion; trigeminal ganglion.

ganglion of cranial nerve 7   figs. 117, 124, 135, 183, 184, 192, 203, 204, 207, 224, 227, 230, 231, 234, 239, 244
also see cranial n. 7; geniculate ganglion
Ganglion arising from preotic neural crest and from the epibranchial placode above the first branchial cleft, also from the dorsolateral placodes in lower vertebrates; forms beside the ganglion of the 8th cranial nerve and supplies sensory fibers to the 7th cranial nerve. *Syn.*: geniculate ganglion.

ganglion of cranial nerve 8   figs. 124, 183, 184, 192, 203, 204, 207, 209, 211, 224, 227, 230, 231
also see acoustic ganglion
Ganglion formed by aggregating cells detached from the otic placode and from the otic vesicle; lies between the geniculate ganglion and the otic vesicle; later divides into the spiral and vestibular ganglia of the inner ear; supplies the fibers of the 8th cranial nerve. *Syn.*: acoustic ganglion.

ganglion of cranial nerve 9   figs. 117, 125, 136, 137, 183, 184, 193, 194, 203, 204, 207, 209, 211, 224, 227, 239
also see cranial n. 9; superior ganglion; petrosal ganglion
Ganglion formed from postotic neural crest and cells from the epibranchial placode above the 2nd branchial cleft; also from the dorsolateral placodes in lower vertebrates; divides subsequently into the superior and petrosol ganglia and

g

supplies sensory fibers to the 9th cranial nerve.

ganglion of cranial nerve 10   figs. 125, 194, 195, 204, 207, 209, 212, 239
also see accessory ganglia; cranial n. 10; jugular ganglion: nodose ganglion
Ganglion formed from postotic neural crest and cells from the epibranchial placode above the 3rd branchial cleft, also from the dorsolateral placodes in lower vertebrates; divides subsequently into the jugular and nodose ganglia and supplies sensory fibers to the 10th cranial nerve.

Gasserian ganglion   *see* ganglion of cranial nerve 5

gastrocoel   *see* archenteron

gastrula   figs. 35–37, 86, 87
The embryonic stage during which the primitive gut or archenteron is formed; the period following the blastula stage and during which extensive cell migrations form the primary germ layers.

geniculate ganglion   figs. 235, 243, 244
see ganglion of cranial nerve 7

genital ridge   figs. 219, 237, 239, 265, 266, 269
A thickening of splanchnic mesoderm (germinal epithelium) and of the underlying mesenchyme on the medial edge of the mesonephros; in the early stages, contains large primordial germ cells; forms the testis or ovary (except that in female birds the right genital ridge fails to develop.)
Syn.: germinal ridge.

genital tubercle   figs. 231, 234, 238, 263
An elevation on the ventral body surface of mammals anterior to the cloacal membrane; enlarges into the phallus and eventually forms the penis of males and clitoris of females.

germinal epithelium   fig. 7
The epithelial covering of the adult ovary and the embryonic gonad; derived from splanchnic mesoderm and primordial germ cells.

germinal vesicle   figs. 16, 24, 27, 43, 79, 144
The much enlarged nucleus of the oocyte; during the prophase of the 1st maturation division, its membrane breaks, releasing most of its contents into the cytoplasm.

germ wall   figs. 149, 150
In bird embryos the outer circular zone surrounding the cellular blastoderm; a region of nuclear proliferation and cell organization which contributes cells to the margin of the epiblast and hypoblast.

gill   figs. 125, 126, 135, 136
also see external gill
In amphibians, filamentous respiratory organs growing out of branchial arches 3 to 6; external gills arise first and then are covered by an operculum; these degenerate as they are replaced by internal gills which subsequently also degenerate during metamorphosis.

gill bars and slits   fig. 56
also see pharynx
In Amphioxus, the gill bars form a series of paired skeletal rods in the pharyngeal wall with gill slits between; contain aortic arches and support the nephridia; arise from the pharyngeal rudiment of the gut.

gill pouch   *see* pharyngeal pouch

gill slit   *see* branchial cleft

glandular oviduct   fig. 145
also see albumen
A large segment of the oviduct of birds located between the isthmus and the uterus; secretes the albumen of the egg when under the stimulus of estrogen.
Syn.: magnum.

glossopharyngeal nerve   *see* cranial nerve 9

glottis   figs. 136, 249, 250
The opening from the pharynx into the trachea of early embryos or into the larynx of later embryos; in mammals, acquires lateral borders (the arytenoid folds) from the 4th and 5th branchial arches.

Graafian follicle   figs. 7, 9
The ovarian follicle of mammals containing a follicular cavity; derive from a primary follicle and either atrophy (atresia) or ovulate to form a corpus luteum.
Syn.: vesicular follicle.

greater omentum   figs. 259, 262
A sac-like membrane attached to the greater curvature of the stomach of birds and mammals; contains a cavity, the omental bursa; derives from the dorsal mesentery of the stomach.

gut   figs. 75, 161, 260
also see archenteron
The digestive tract; derives from the archenteron.

head ectoderm   figs. 93, 121, 159, 163–165, 166, 173, 174, 175
also see epibranchial placodes; lens placode; olfactory placodes; otic placode
The epithelial outer covering of the head; mostly forms epidermis but some placodes (thickenings) arise which contribute to the sense organs and cranial ganglia.

head fold   figs. 154, 164, 176, 177
A downward bend of membranes around the head which mark the boundaries of the embryonic area; undercuts the head, separating it from the extraembryonic area to form the subcephalic pocket; forms by invagination, the ventral surface of the head and the foregut and is posteriorly continuous with the body folds.

head mesenchyme   figs. 93, 96–98, 106, 110, 121–123, 155, 159, 164, 165, 175–178, 189, 190, 193, 211
also see dense mesenchyme
A loose tissue surrounding the brain and foregut, mostly of mesodermal origin; derives from the paraxial mesoderm anterior to the somites, the prechordal plate, and the neural crest; forms the following head structures: blood vessels, skull, head muscles, and connective tissue.

head organizer   fig. 91
Inductor of head parts; consists of foregut roof, prechordal plate and anterior notochord; the first area to pass over the dorsal lip of the blastopore during gastrulation in amphibians. *Syn.*: head inductor.

head plexus   figs. 160, 173, 186
A dense capillary network surrounding the brain anterior to the myelencephalon; supplied by the internal carotid arteries and drained by the anterior cardinal veins; some head vessels develop later from the plexus.

head process   *see* notochordal process

heart   figs. 103, 104, 111, 115, 116, 130, 131, 230, 233
also see atrium; bulbus cordis; sinus venosus; ventricle
In early stages, a tubular organ divided by constrictions into sinus venosus, atrium, ventricle, and bulbus cordis; its wall is formed of an inner endocardial layer and an outer epimyocardium; arises from paired heart tubes derived from splanchnic mesoderm which fuse beneath the foregut; in air-breathing vertebrates, becomes more or less completely divided longitudinally in the later stages to provide a separate pulmonary circulation.

heart mesoderm   figs. 101, 159, 164
also see heart
The heart-forming tissue; in amphibians from paired areas of the mesodermal mantle passing over the lateral lips of the blastopore, moving forward then turning ventrally to a position under the foregut; in birds it arises from splanchnic mesoderm of both amnio-cardiac vesicles adjacent to the anterior intestinal portal.

Hensen's node   *see* primitive knot

hepatic caecum   fig. 56
In Amphioxus, a blind outgrowth of the stomach extending anteriorly along the right side of the pharynx, probably serves as a digestive gland; arises in a late larval stage. *Syn.*: liver, midgut caecum.

hepatic ducts   fig. 264
A series of small ducts conducting bile from the liver to the common bile duct; arise from the hepatic diverticulum.

hepatic portal vein   figs. 235, 260, 263, 265
The vessel which carries blood from the superior mesenteric and splenic veins to the ductus venosus and sinusoids of the liver; derives from parts of the right and left vitelline veins.

hepatic vein   figs. 231, 260
also see vitelline veins
The trunk vein of the liver; drains initially into the sinus venosus and later into the inferior vena cava; arises from the anterior segment of the vitelline veins.

hindbrain   *see* rhombencephalon

hindgut   figs. 100–102, 104, 106, 107, 114, 115, 118, 119, 131, 139, 202, 208, 221, 222, 231
also see cloaca

The posterior part of the embryonic gut; extends from the midgut to the tail in amniotes; formed as the tail fold passes under the posterior trunk region; forms successively tail gut, cloaca, colon, and posterior small intestine; in amphibians, forms the rectum.

hindleg bud   figs. 140, 207, 230, 233, 234, 267, 269
also see foreleg bud; leg bud.
The rudiment of the hindleg; arises as a thickening of somatic mesoderm in the body wall; an ectodermal thickening, the apical ridge, forms on its margin in amniotes; homologous to the leg bud of birds and humans.

Hofbauer cell   fig. 275
Large, vacuolated spherical cells within the stroma of chorionic villi; function uncertain but may be phagocytes.

hyoid arch   figs. 102, 106, 107, 118, 119, 193, 194, 203, 214, 224, 228, 230, 233, 239, 247–250
also see aortic arch 2
The second branchial arch; arises as a thickening of the pharyngeal wall between the first and second branchial clefts; contains the second aortic arch in the early stages, forms the stapes (columella), styloid process, stylohyoid ligament, part of the hyoid bone, root of the tongue, and the facial muscles. *Syn.*: branchial arch 2.

hyomandibular cleft   *see* branchial cleft 1

hypoblast   figs. 149, 150, 152, 155–158
The inner or lower layer of the blastoderm of birds and mammals; lies between the epiblast and subgerminal cavity or between epiblast and yolk; formed by inward migration and aggregation of large yolky cells; mostly forms endoderm.

hypochordal rod   *see* subnotochordal rod

hypoglossal nerve   *see* cranial nerve 12

hypomere   *see* lateral plate mesoderm

hypophysis   figs. 96, 101, 104, 109, 116, 118, 120
also see infundibulum; Rathke's pouch
An endocrine gland beneath the hypothalamus; derives from the infundibulum and Rathke's pouch or, in amphibians, from infundibulum

and a solid ingrowth from stomodeum. *Syn.*: pituitary.

iliocolon ring   figs. 56, 71, 73
In Amphioxus, a thickened heavily ciliated segment of the intestine.

immature sperm   figs. 15, 141
also see differentiating spermatid
The sperm during its final stage of differentiation; some further elongation of the head with its dense nucleus and of the tail will occur.

inferior ganglion   *see* petrosal ganglion

inferior vena cava   *see* posterior vena cava

infundibulum   figs. 101, 102, 104, 106, 110, 116, 120, 131, 170–174, 183–185, 187, 192, 204, 207, 208, 214, 224, 226, 232, 234, 237
also see hypophysis
A ventral evagination of the prosencephalon; lies in the floor of the diencephalon and later in the hypothalamus; subsequently evaginates the neural (posterior) lobe of the hypophysis.

interatrial foramen   figs. 231, 252
also see interatrial foramen I; atrial septum
An opening in the atrial septum allowing blood to pass from the right to the left side of the heart in tetrapods; closes soon after breathing begins, completing longitudinal division of the heart.

interatrial foramen I   fig. 254
also see atrial septum I
An opening in the atrial septum I allowing blood to pass from the right to left atria; as the foramen I closes near the endocardial cushion a new interatrial foramen II opens through the anterior region of the atrial septum I allowing continued flow between the atria.

interatrial foramen II   fig. 254
also see atrial septum I
A second opening in the atrial septum I allowing blood to pass from the right to the left atria after closure of interatrial foramen I.

intermediate mesoderm   *see* nephrotome

internal carotid arteries   figs. 186, 206, 207, 212–214, 225, 227, 235, 236, 244, 245
The main arterial supply to the brain; arise as anterior extensions of the dorsal

aortae; later acquire additions from the dorsal aortae and 3rd aortic arches.

internal nares   fig. 130
also see external nares; olfactory pit
The inner opening of the nasal passages; formed when the olfactory pits extend downward and break through the roof of the mouth.

interphase   fig. 143
The period in the cell cycle between mitoses; in proliferating tissues a time of synthesis of DNA, RNA, and protein.

intersegmental arteries   figs. 206, 236, 248, 270
Originally small paired branches of the dorsal aortae arising between the somites; contribute to the following arteries: vertebrals, subclavians, intercostals, and lumbars.

intersomitic grooves   fig. 159
also see somites
The spaces separating somites; the first to form lies between the 1st and 2nd somites; the others form within the segmental mesoderm in an anterio-posterior sequence; subsequently obliterated by fusion of adjacent somites.

interstitial cells   figs. 1, 2, 5, 7, 77
In the testis, clusters of endocrine gland cells (Leydig cells) between the seminiferous tubules which secrete testosterone; in the ovary, clusters of gland cells scattered in the cortex; derive from atretic follicles.

interventricular foramen   figs. 231, 253, 254
also see ventricular septum
An opening in the ventricular septum allowing blood to cross between the right and left ventricles in tetrapods; closes during division of the bulbus cordis and the atrioventricular canal.

interventricular sulcus   fig. 253
A longitudinal groove on the surface of the primitive ventricle marking the plane of its impending division into right and left ventricles.

intervillus space   figs. 273, 274
also see placental villi

In the placenta, region filled with maternal blood and in which the placental villi are suspended; maternal blood supply to the intervillus space is by way of the spiral arteries of the basal plate; veins of the basal plate return the blood to the maternal circulation.

intestinal loop   figs. 231, 234, 237, 238
also see caecum; colon, small intestine
A ventral extension of the intestine into the umbilical cord of mammals bearing an enlargement, the caecum; coiling soon retracts it into the peritoneal cavity.

intestine   figs. 38–42, 46, 55, 56, 69–71, 120, 130, 131, 135, 137–139, 232, 237
also see caecum; colon; duodenum; intestinal loop; rectum
Segment of gut following the stomach; derives from both foregut and hindgut in amniotes, from the midgut as well in amphibians, and from archenteron in Amphioxus, sea urchin and starfish; the intestinal lining develops from gut endoderm; muscle, connective tissue, blood vessels, and serosa develop from splanchnic mesoderm.

isthmus   figs. 145, 184, 187, 203, 204, 207, 226, 232, 237, 240
also see glandular oviduct
A narrow part of the oviduct of birds where the shell membranes of the egg are formed; the narrowed connection between the cerebral aquaduct of the mesencephalon and the 4th ventricle of the metencephalon; thickening of the floor of the brain progressively reduces the isthmus.

jugular ganglion   figs. 231, 235, 242
also see ganglion of cranial nerve 10
A large ganglion of the 10th cranial nerve dorsal to the nodose ganglion; contributes sensory fibers to the 10th nerve

junctional complex   fig. 6
An intercellular organelle of epithelial cells which binds the cells together and forms a barrier against passage of substances between cells; consists of the zonula adherens, zonula occuludens, and gap juctions.

lampbrush chromosomes   fig. 79
Enlarged chromosomes of oocytes with

many lateral loops which extend the length of the chromosome during the period of intense RNA transcription and yolk synthesis.

laryngotracheal groove   figs. 187, 196, 208, 214

A trough in the floor of the posterior pharynx from which arise the lung buds; also contributes to the larynx and trachea.

larynx   fig. 251

The voice box; derives from upper trachea, the floor of the pharynx, and branchial arches 4 and 5.

lateral body fold   figs. 188, 200, 201, 218, 219

A depressed fold in the somatopleure beside the embryonic trunk; together with the head and tail folds, to which it connects, forms the boundary between the embryonic and extraembryonic regions.

lateral bronchus   fig. 256

A secondary branch of the respiratory tract lateral and anterior to the stem bronchus.

lateral lip   figs. 89, 91

also see dorsal lip; ventral lip
The lips on the right and left rims of the blastopore formed by extensions of the dorsal lip; in amphibians, midgut endoderm, lateral plate and heart mesoderm pass over the lateral lips during gastrulation.

lateral nasal process   fig. 251

An elevation on the embryonic face lateral to the olfactory pit; fuses with the maxillary and median nasal processes and forms the sides of the nose.

lateral plate mesoderm   figs. 91, 94, 95, 99, 102, 112, 113, 127, 166, 167, 172, 185

also see ventral mesoderm; somatic mesoderm; splanchnic mesoderm
The mesodermal layer lateral and ventral to the nephrotome; split by the coelom into somatic and splanchnic mesoderm. *Syn.*: hypomere.

lateral swellings   figs. 246-248

also see tongue
Paired elevations on the mandibular process within the mouth; fuse with tuberculum impar to form the body of the tongue.

lateral transverse vein   fig. 236

In the pig embryo, part of a plexus of small veins draining the mesonephros.

lateral ventricles   figs. 216, 251

also see third ventricle; telencephalon
Lateral extensions of the original third ventricle of the telencephalon into the cerebral hemispheres; retain connections to the third ventricle of the diencephalon through the foramen of Monroe; the thin roof projects into the lumen as a choroid plexus.

left atrium   figs. 251–253, 255

also see atrial septum; atrium
The left division of the primitive heart atrium in tetrapods; separated from the right atrium by the atrial septum; receives blood through the interatrial foramen and the pulmonary veins before breathing begins; after closure of interatrial foramen, supplied by increased pulmonary flow. *Syn.*: left auricle.

left auricle   *see* left atrium

left diverticulum   figs. 69, 70, 73

also see dorsal diverticulum; preoral pit
A ventral thick-walled extension of the dorsal diverticulum; an external opening, the preoral pit, is near the mouth.

left horn of sinus venosus   figs. 253, 254

also see sinus venosus

The part receiving the left common cardinal, left vitelline, and left umbilical veins; conducts the flow to the transverse sinus venosus whence it passes into the right atrium through the sinoatrial opening; as the venous return is directed to the right horn of the sinus venosus, the left horn is reduced, forming finally the oblique vein of the left atrium.

left umbilical vein   *see* umbilical vein.

left ventricle   figs. 252, 253, 255, 256

also see ventricle
Heart chamber formed from the partitioning of the primitive ventricle by the ventricular septum; receives blood from the left atrium and delivers it under high pressure to the ascending aorta.

leg bud figs. 203, 208, 209, 221, 222, 224–226, 228, 239
also see foreleg bud; hindleg bud
The rudiment of the leg; arises as a thickening of somatic mesoderm of the body wall, later bearing an ectodermal thickening, the apical ridge.

lens figs. 130, 134
also see lens placode; lens vesicle
Derived from the lens vesicle by a thickening of the inner wall of the vesicle accomplished by elongation of the epithelial cells there; eventually enclosed by a lens capsule.

lens placode figs. 83, 103, 109, 174
also see lens vesicle
A thickening of head ectoderm overlying the optic vesicle; invaginates to form the lens vesicle and subsequently the eye lens.

lens vesicle figs. 117, 122, 183–185, 193, 203, 204, 207, 215, 248
also see lens placode
An ectodermal sac within the optic cup; derives from lens placode, forms the eye lens.

leptotene stage figs. 3, 12, 143
also see prochromosome stage
An early stage of the first maturation division in spermatogenesis and oogenesis; chromosomes have the form of separate long thin threads except that the X-chromosome may be a dense contracted body; followed by the zygotene stage; during which synapsis or pairing of homologous chromosomes occurs.

lesser omentum figs. 259, 262
A membrane which attaches the lesser curvature of the stomach to the liver; derives from the ventral mesentery of the stomach.

lip figs. 130–133
Extensions of skin on the upper and lower jaws of amphibian tadpoles forming the border of the mouth.

liver figs. 120, 130, 137, 138, 207, 208, 224, 226, 227, 230, 232–234, 237–239, 257, 260, 265
also see liver diverticulum; liver sinusoid; anterior liver diverticulum; posterior liver diverticulum
The largest of the digestive glands;

important in fetal life for blood homeostasis and blood formation; arises as a ventral diverticulum of the foregut in amphibians and mammals, and as two buds on the duodenum in birds; the buds branch and anastomose around the ductus venosus.

liver diverticulum figs. 90, 94, 99, 101, 103, 104, 107, 112, 115, 120, 126
also see anterior liver diverticulum; hepatic ducts; liver; posterior liver diverticulum
The rudiment of the liver, gall bladder, hepatic ducts, and common bile duct; arises as a ventral evagination of the foregut in amphibians and mammals, and as two buds on the duodenum of birds.

liver sinusoids fig. 259
The smallest blood vessels of the liver; differ from capillaries in that their walls contain phagocytes; derive originally from the ductus venosus and vitelline veins.

lumbosacral plexus fig. 270
An anastomosis of spinal nerves in the posterior trunk region which supplies the nerves of the hind limb.

lungs figs. 137, 138, 235
also see apical bronchus; laryngotracheal groove; lateral bronchus; lung bud; stem bronchus
Arise as a ventral diverticulum of the pharynx which branches repeatedly to form the endodermal lining of the trachea, bronchi, and lungs; the mesoesophagus and lining of the pleural cavities form the mesodermal parts—muscle, connective tissue, and pleura; the pulmonary arteries arise from the 6th aortic arch and invade the lung; the pulmonary veins grow in from the atrium.

lung buds figs. 120, 207, 210, 215, 227, 232, 239
also see laryngotracheal groove; lateral bronchus; lungs; stem bronchus
The paired rudiments of the lungs and bronchi; arise from the laryngotracheal groove of the pharynx. *Syn.*: primary bronchi.

lymph sinus figs. 79, 134
A large, lymph-filled space; drains into the veins; the cavity of the hollow amphibian ovary.

m

macromeres figs. 32, 63, 64, 84, 85, 89
The largest blastomeres or cells formed during cleavage; form endoderm and in the sea urchin ventral ectoderm also; in Amphioxus, endoderm, notochord and mesoderm.

mammary ridge figs. 262, 263, 265, 266
The rudiment of the mammary gland; initially appears as an ectodermal thickening extending longitudinally between the bases of the limb buds; at the site of the mammary glands, the mammary ridges form tubular ingrowths, the milk ducts; renewed development of the mammary glands at puberty in females is a response to the ovarian hormones; development completed during pregnancy.

mandible figs. 224, 226, 234, 237–239
also see mandibular arch; mandibular process
The lower jaw; formed by the fusion of the mandibular processes of the first branchial arch.

mandibular arch figs. 107, 109, 118, 119, 122, 123, 183, 184, 193, 212, 228, 230, 232, 247
also see aortic arch 1; mandibular process; maxillary process
The most anterior branchial arch, composed of a mandibular process forming the posterior border of the stomodeum, and a maxillary process anterior to the stomodeum; in the early stages it contains the first aortic arch; its posterior boundary is the 1st branchial cleft; forms the lower jaw and sides of the upper jaw. *Syn.*: branchial arch 1.

mandibular process figs. 203, 204, 208–210, 213, 214, 227, 228, 233, 248, 249
also see mandible; mandibular arch
The posterior division of the mandibular arch; forms the mandible, Meckel's cartilage, body of tongue, malleus, salivary glands, jaw muscles.

mandibular ramus of cranial nerve 5 figs. 245, 246
also see cranial nerve 5
The posterior division of the fifth cranial nerve; innervates the mandible and jaw muscles.

mantle layer figs. 241, 249, 259, 261
The middle layer of the developing neural tube; contains neuroblasts from the ependymal layer and their nerve fibers; forms the gray matter.

marginal layer figs. 137, 261
The outer layer of the developing neural tube; contains neuroblasts from the inner layers and nerve fibers; forms the white matter.

marginal zone figs. 85, 86, 88, 89
The part of the animal hemisphere nearest the vegetal hemisphere; is turned in during gastrulation to form mesoderm and foregut endoderm.

maternal blood fig. 274
also see intervillus space
In the living placenta, maternal blood fills the intervillus spaces; mostly drained out in microscopic preparations of placenta; separated from fetal blood by placental barrier.

maturation division I figs. 2, 10, 13, 17, 57, 77, 78, 142
The first of two specialized cell divisions during the formation of sperm and eggs; the prophase is long and includes synapsis, chromosome replication, and crossing over; reduces the chromosome number to the haploid state; forms the secondary spermatocytes of males and secondary oocyte and a polar body in females. *Syn.*: meiosis I, reduction division I.

maturation division II figs. 10, 13, 18, 78
The second of two specialized cell divisions during the formation of sperm and eggs; begins immediately after the first maturation division but in many eggs is not completed until after fertilization; no chromosome replication occurs; forms the spermatids of males and the second polar body and mature egg in females. *Syn.*: reduction division II, meiosis II.

mature ova fig. 24
also see oocyte; primary oocyte; egg pronucleus
The female germ cells after completion of the maturation divisions; derived from oocytes; in sea urchins, found in the lumen of the ovary where they may be fertilized.

maxillary process   figs. 203, 204, 209, 210, 213, 214, 227, 228, 234, 248–250
also see mandibular arch
The anterior division of the mandibular arch; forms the cheek, lateral part of upper jaw, palate, incus.

maxillary ramus of cranial nerve 5
fig. 246
also see cranial nerve 5
The middle division of the 5th cranial nerve; innervates the upper jaw and face.

medial transverse vein   fig. 236
In the pig embryo, part of a plexus of small veins draining the mesonephros.

median nasal processes   fig. 251
Elevations on the embryonic face medial to the olfactory pit; the two median nasal processes fuse to form medial part of the upper jaw.

medulla   figs. 7, 9
The inner or deep division of an organ; in the ovary, a region of connective tissue and blood vessels lacking follicles.

medullary groove   *see* neural groove

medullary plate   *see* neural plate

meiosis   *see* maturation division

mesencephalon   figs. 96, 101, 103, 104, 109, 115, 116, 122, 170, 171, 173, 175, 183–185, 187, 189, 203, 204, 208–214, 224–226, 228, 230, 232–234, 237–243

The middle primary vesicle of the brain; forms visual and auditory centers (corpora quadrigemina in mammals) and motor centers for movements of the head; its cavity narrows to form the cerebral aqueduct; bears the 3rd and 4th cranial nerves. *Syn.*: midbrain.

mesenchyme   fig. 52
also see head mesenchyme; primary mesenchyme; secondary mesenchyme
Loosely scattered cells which in early development spread through the blastocoel; may derive from any germ layer but in sea urchins arises from micromeres and archenteron, and in starfish from the archenteric vesicle; in these groups, constitute part of the early mesoderm and forms skeleton.

mesentery   figs. 267, 268

A supporting membrane attached to organs within the coelom; carries the vascular and nerve supply of organs; formed by fusion of two layers of splanchnic mesoderm.

mesoderm   figs. 75, 91, 98, 120, 154, 156–158, 162, 181, 188
also see primary mesenchyme; secondary mesenchyme, coelomic vesicles; head mesenchyme; lateral plate mesoderm, dorsal mesoderm; segmental mesoderm; somite; nephrotome; ventral mesoderm
The middle primary germ layer; formed mostly by the gastrulation movements; forms dermis, muscle, skeleton, blood vessels, blood (excepting possibly lymphocytes), connective tissues, kidneys, ureters, gonads (excepting germ cells), reproductive tracts, peritoneum.

mesodermal crescent   figs. 72, 74
also see chorda-neural crescent
A light zone visible in the egg and during cleavage of Amphioxus embryos located on the ventral side of the vegetal hemisphere; during gastrulation it divides and migrates dorsally to positions on either side of the notochord, then separates from the archenteron wall to form the somites, which constitute the entire mesoderm in Amphioxus.

mesodermal groove   fig. 72
also see enterocoel
A dorsal extension of the archenteron formed by the evagination of the somites; will form the enterocoel.

mesomeres   fig. 32
also see nephrotome
The cells of intermediate size that compose the animal hemisphere of sea urchin embryos during cleavage; form the ectoderm of the gastrula and larva. Also the nephrotome of vertebrates.

mesonephric duct   figs. 139, 200, 201, 219, 221–223, 228, 235, 237, 239, 265–269
The excretory duct of the mesonephros; formed initially as the pronephric duct by the caudal growth of the pronephric buds to the cloaca; contributes to the metanephros of amniotes by forming one of its rudiments, the ureteric bud; mostly degenerates in female amniotes but in males forms the ductus epididymis,

ductus deferens, and seminal vesicles;
forms the duct of the adult kidney
(opisthonephros) of amphibians.
Syn.: archinephric duct; Wolffian duct.

mesonephric glomeruli   figs. 231, 235–239,
263, 265, 269

Tufts of capillaries within Bowman's
capsules which together form the renal
corpuscles of the mesonephros; similar
structures form in the metanephros;
glomeruli of the pronephros are associated
with the coelom rather than with
Bowman's capsule.

mesonephric ridge   fig. 219
also see mesonephros
A bulge or thickening extending into the
dorsal part of the embryonic coelom at
midtrunk levels; formed by the growing
mesonephros.

mesonephric tubules   figs. 184, 219, 227,
230, 235–239, 261, 263, 265, 269

The kidney tubules of the mesonephros;
possess glomeruli and a well-formed,
coiled tubular structure; excretory during
the embryonic period of amniotes; most
degenerate but some form the efferent /
ductules of male amniotes; forms adult
kidney tubules in amphibians.

mesonephros   figs. 139, 203, 208, 220,
226, 232–234, 259 262, 266
also see mesonephric tubules;
mesonephric glomeruli; mesonephric
ridge; mesonephric duct
The second or middle kidney of amniotes;
contains well-formed tubules with
glomeruli that produce urine during the
embryonic period; the arrangement of the
tubules is not segmental; the pronephric
duct is appropriated as the mesonephric
duct; mostly degenerates in the adult
amniote except that in males the caudal
parts form the male reproductive tract
(efferent ductules, ductus epididymis,
ductus deferens, seminal vesicles).
Syn.: Wolffian body.

mesovarium   fig. 9
The mesentery of the ovary; provides
support for the ovary.

metanephric diverticulum   fig. 235
also see pelvis of metanephros; ureter
A rudiment of the metanephros; arises as

an outgrowth of the mesonephric duct
near its junction with the cloaca; the stalk
becomes the ureter and the expanded
distal end forms the pelvis and collectings
ducts of the metanephros.
Syn.: ureteric bud.

metanephric duct   *see* ureter

metanephrogenic mesenchyme   fig. 268
A strand of dense mesenchyme
surrounding the pelvis of the metanephros;
continuous anteriorly with the
mesonephrogenic tissue; derives from
the caudal nephrotomes; forms the
metanephric tubules. *Syn.*:
metanephrogenous tissue.

metanephrogenous tissue
see metanephrogenic mesenchyme

metanephros   fig. 268
also see metanephric diverticulum;
metanephrogenic mesenchyme; pelvis
of metanephros; ureter
The last of three pairs of kidneys to
form in amniotes; metanephric tubules
arise from the posterior end of the
nephrogenic cord (metanephrogenic
mesenchyme); the metanephric duct
and pelvis are formed from the
metanephric diverticulum (ureteric
bud), an outgrowth of the mesonephric
duct; begins functioning in the embryo
and continues as the adult kidneys.

metencephalon   figs. 131, 183–185,
187, 189, 203, 204, 207–211, 224–228,
230, 232–234, 237–244

The anterior division of the
rhombencephalon; its roof expands
greatly to form the cerebellum while
the pons forms in its floor; nerve
centers (nuclei) for several cranial
nerves develop within, including those
of the 5th, 6th, 7th, and 8th; its
cavity becomes the 4th ventricle of
the brain.

micromeres   figs. 32, 33, 63, 64, 84, 85
The smallest cells of the cleavage stages;
lie near the vegetal pole in the sea urchin
and migrate into the blastocoel to form
the primary mesenchyme; compose the
entire animal hemisphere of amphibian
embryos, forming ectoderm and mesoderm
after gastrulation; compose the animal

hemisphere of Amphioxus where they form ectoderm.

midbrain   *see* mesencephalon

midgut   figs. 95, 99, 101, 102, 104, 106, 113, 121, 165, 179, 200, 210, 219
also see anterior intestinal portal
In amphibians the middle part of the gut with a small lumen and thick yolky floor; derives from archenteron and will form the small intestine; in amniotes, the middle part of the gut whose floor is the cavity of the yolk sac (yolk-filled in reptiles and birds); it is steadily diminished by the lengthening of the foregut and hindgut to a mere yolk stalk attached to the small intestine.

mitochondria   fig. 11
Filamentous or globular membranous organelles found in all eukaryotic cells; have characteristic internal membranous folds or cristae; contain enzymes of the Krebs cycle which transfer energy released by oxidative phosphorylation to ATP. *Syn.:* chondriosomes.

mitochondrial body   fig. 11
A fused mass of mitochondria found in the spermatids of some insects.
Syn.: nebenkern.

mitotic figure   fig. 84
also see cleavage division
The mitotic apparatus, consisting of chromosomes, spindle fibers, and centrioles; appears during each cleavage division; often indistinct in preparations used for class work.

motor root   *see* ventral root of spinal nerve

mouth   figs. 38–42, 54, 55, 130, 224, 226, 237, 246–250
also see oral evagination; stomodeum

The anterior opening of the digestive tract; derived partly from an ectodermal invagination on the ventral side of the head, the stomodeum; the endodermal rudiment arises from the anterior wall of the foregut and is for a time separated from the stomodeum by the oral plate; rupture of the membrane opens the mouth.

muscle   fig. 134
also see myotome

Skeletal muscles of vertebrates mostly arise from myotomes but some head muscles (those of the jaws and eyes) arise from head mesenchyme and limb muscles from lateral plate (in fishes fin muscles arise from myotomes).

muscle fibrillae   fig. 75
also see myotomes
Contractile filaments in developing muscle cells of the myotome.

myelencephalon   figs. 131, 135–138, 183–185, 187, 189–195, 203, 204, 207, 208, 210, 211, 224–228, 230, 232–234, 237, 238, 240–245

The posterior division of the rhombencephalon and the most posterior part of the brain; a transition region from brain to spinal cord; contains nerve centers (nuclei) of cranial nerves 9 to 12; cranial nerves 9 and 10 are attached to its sides and in amniotes, nerves 11 and 12 as well; its cavity becomes part of the 4th ventricle of the brain and its roof, the choroid plexus; forms the medulla.

myometrium   fig. 272
also see endometrium; smooth muscle
The muscle layer of the uterus; composed of bands of smooth muscle and numerous blood vessels; myometrial contractions transport sperm into uterine tubes and at the end of pregnancy produce the labor contractions.

myotome   figs. 56, 126, 131, 139, 199, 215, 248, 262, 269–271
also see somites
The division of the somite which forms skeletal muscle of the body wall; lies at first in the dorsal region of the somite then migrates ventrally under the dermatome; its cells elongate longitudinally forming a muscle segment; may also contribute to limb musculature; myotomes are comparatively large in amphibian embryos.

nasal cavity   fig. 120
also see olfactory pits
A canal extending from the nostril to the mouth; formed from the olfactory pits as they deepen and break through the roof of the mouth; forms the olfactory organ.
Syn.: nasal passage.

n

nasal pit   *see* olfactory pits

nasal placode   *see* olfactory placodes

nephrotome   figs. 99, 179, 201
also see pronephros
Stalk-like connection between the somite and lateral plate; forms segmental buds in the anterior region which hollow out to form the pronephric tubules and ducts; at posterior levels, forms mesenchyme which develops into tubules of the mesonephros and metanephros; contributes to the gonads. *Syn.*: mesomere; intermediate mesoderm.

nervous layer of ecoderm   fig. 87
The inner layer of ectoderm, covered by the epithelial layer; thickens in the regions of the neural plate and ectodermal placodes.

nervous layer of optic cup   figs. 122, 134, 215, 248
also see optic cup
The inner layer of the optic cup; arises from the lateral wall of the optic vesicle; forms the nervous or sensory layer of the retina and the optic nerve fibers.

neural canal   fig. 75
also see neural tube; neurocoel
The cavity of the neural tube; in many embryos temporarily opens externally through a neuropore and connects with the archenteron through the neurentric canal. *Syn.*: central canal; neurocoel.

neural crest   figs. 93, 94, 98, 105, 109–111, 164, 174, 178
also see ganglion of cranial nerve 5, etc.; spinal ganglia
An ectodermal mesenchyme arising from the neural folds; aggregates in many locations to form cranial ganglia, spinal ganglia, autonomic ganglia and adrenal medulla; attaches to epidermis to form pigment cells and to the neural tube to form the pia mater; forms the neurolemma sheath cells of nerves; some migrate into the branchial arches to form the visceral skeleton.

neural ectoderm   fig. 74
also see chorda-neural crescent; neural crest; neural plate

That part of the ectoderm which normally develops into neural structures or tissue.

neural folds   figs. 74, 75, 90, 93–95, 159, 160, 163, 165, 166, 170
The elevated edges of the neural plate; opposite neural folds are brought together by bending and closing of the neural groove; fuse and bud off streams of neural crest cells.

neural groove   figs. 72, 73, 89, 90, 95, 163, 165, 169, 181
also see neural plate; neural folds; neural tube
A trough formed by the bending or rolling up of the neural plate; closes to form neural tube. *Syn.*: medullary groove.

neural plate   figs. 72, 75, 93–95, 154–157, 165–167, 169
also see neural folds; neural groove; neural tube
A thickening of the dorsal ectoderm caused by its transverse contraction; the placode or rudiment of the central nervous system. *Syn.*: medullary plate.

neural tube   figs. 56, 68–71, 75, 159, 164, 169, 172, 173, 185, 207
also see mesencephalon; neural groove; neural plate; prosencephalon rhombencephalon; spinal cord
The rudiment of the central nervous system; formed by bending and closing of the neural plate; the anterior large end forms the brain, the posterior part, the spinal cord; in amphibians the extreme posterior end contributes to the tail somites.

neurenteric canal   figs. 72, 74, 100
A temporary connection between the caudal end of the neural groove or neural tube and the archenteron or yolk sac cavity; occurs in many embryos, including frog and man.

neurocoel   fig. 149
The lumen or cavity of the neural tube. *Syn.*: neural canal.

neuromere   figs. 185, 211
A minor segment of the brain formed by transverse constrictions; particularly prominent in the myelencephalon; later vanishes.

neuropore   fig. 73
also see neural canal

The temporary anterior external opening of the neural canal.

nodose ganglion   figs. 235, 246
also see ganglion of cranial nerve 10
A large ganglion of the 10th cranial nerve lying ventral to the jugular ganglion; contributes sensory fibers to the 10th nerve.

notochord   figs. 56, 69-73, 90, 91, 94, 95, 97-101, 104, 105, 110-120, 123-125, 127-131, 135-140, 159, 161, 164, 167, 169, 172, 173, 175, 180, 184, 185, 187, 191-193, 204, 207, 208, 212, 213, 220, 226, 227, 235, 237, 245, 270
also see notochordal process
The axial skeleton of chordate embryos; arises in amphibians and Amphioxus from cells turned in over the dorsal lip of the blastopore, and in amniotes from cells extending anteriorly from the primitive knot; shows marked elongation and in amphibians extending the embryonic axis; lies under the neural tube from the mesencephalon to the end of the spinal cord; its gelatinous cells acquire a tough sheath forming a flexible skeletal rod; in higher vertebrates it is small and mostly replaced by the vertebral column; in Amphioxus extends anterior to the brain. *Syn.*: chorda.

notochordal process   figs. 154, 155
also see notochord
In amniotes, a band of mesodermal cells extending anteriorly from the primitive knot; the rudiment of the notochord. *Syn.*: head process.

nuclear envelope   *see* nuclear membrane

nuclear membrane   figs. 27, 43, 79
Encloses the substance of the nucleus; at the ultrastructural level, it is double-layered with "pores"; destroyed during prophase of mitosis and re-formed during telophase. *Syn.*: nuclear envelope.

nucleolus   figs. 27, 43, 58, 79
One or more dense spherical granules in the nucleus of many cells; composed of ribosomal RNA, (rRNA) and protein; disappears during mitosis to re-form in connection with a pair of nucleolar chromosomes; large in cells with high rates of protein synthesis; multiple

nuceoli occur in amphibian oocytes as a result of rRNA gene amplification.

nucleus   figs. 28, 50, 58, 60, 83, 150
also see germinal vesicle; nuclear membrane; nucleolus; pronucleus
A large body within the cell during interphase; contains the chromosomes and often one or more nucleoli; enclosed by a nuclear membrane; disappears during mitosis; site of synthesis of most of the DNA and RNA of the cell.

oculomotor nerve   *see* cranial nerve 3

olfactory organ   fig. 133
also see nasal cavity; olfactory pits; olfactory placodes. A tubular passage derived from olfactory pits which extends inward opening through the roof of the mouth as internal nares; olfactory neurons arise from part of the organ extending olfactory nerve fibers into the telecephalon. *Syn.*: nasal passage.

olfactory pits   figs. 103, 106, 108, 115, 118, 120, 121, 203, 204, 207, 216, 224, 228, 230, 232-234, 238, 251
also see nasal cavity; olfactory placodes; olfactory organs
Cavities on the lateral surfaces of the head anterior to the eyes; arise by invagination of the olfactory placodes; deepen and break through the roof of the mouth in air breathers to form the nasal cavities; olfactory cells differentiate from their walls as do the olfactory nerves. *Syn.*: nasal pit.

olfactory placodes   figs. 102, 195
also see olfactory pits
Paired ectodermal thickenings on the lateral surfaces of the head anterior to the eyes; invaginate to form the olfactory pits; the rudiments of the nasal passages. *Syn.*: nasal placode.

omental bursa   figs. 259, 262
The cavity of the greater omentum; arises as an invagination into the dorsal mesentery of the stomach; connected with the peritoneal coelom; two bursae form in birds.

omphalomesenteric artery   *see* vitelline arteries

omphalomesenteric veins   *see* vitelline veins

O

oocyte figs. 7, 8, 27, 79

also see primary oocyte

The immature egg, distinguishable from other ovarian cells by gigantic size and prominent nucleus (germinal vesicle); grows from the smaller oogonium; becomes a mature egg upon completion of its growth and two maturation divisions.

opercular chamber figs. 135–137

also see opercular opening

Gill chambers formed by a membranous outgrowth of the hyoid arch of tadpoles, extending posteriorily over the gills and branchial clefts; the two chambers connect ventrally and open through an external spiracle on the left side. *Syn.:* branchial chamber.

opercular opening fig. 120

A persistent excurrent opening at the posterior edge of the operculum on the left side of the larval frog body; closed during metamorphosis by proliferation of tissue which fills the opercular cavity. *Syn.:* spiracle.

ophthalmic ramus of cranial nerve 5 fig. 245

also see cranial nerve 5

The anterior division of the 5th cranial nerve; distributes sensory fibers to the anterior facial region.

optic cup figs. 103, 109, 115, 117, 120, 122, 183–185, 193, 203, 204, 207, 224, 230, 234, 246

also see nervous layer of optic cup; optic vesicle; pigmented layer of optic cup

A double-walled chamber formed by invagination of the optic vesicle; the lens vesicle lies in its "mouth"; remains connected to the diencephalon by the optic stalk; the outer wall forms the pigment epithelium of the retina; the inner wall forms the nervous layers of the retina and the optic nerve fibers which grow through the stalk to the brain; the rim of the cup contributes to the iris and ciliary body; the optic cup of mammals is small. *Syn.:* eye cup.

optic stalk figs. 109, 122, 184, 194, 210, 215, 228, 238, 248

The narrow connection of the optic cup to the diencephalon; guides the growing

optic nerves from the optic cup to the brain.

optic vesicle figs. 96, 102, 105, 106, 170, 173, 174, 204

also see optic cup

A lateral evagination of the prosencephalon; by invagination of outer wall forms the optic cup and subsequently the retina. *Syn.:* eye vesicle.

oral arms figs. 38–41

also see anal arms

Slender, paired extensions of the dorsal body wall of pluteus larvae; supported by skeletal rods and bearing bands of cilia; function to stabilize and propel the larva and to collect food

oral evagination figs. 104, 107, 109, 116, 119

The endodermal rudiment of the mouth; an anterior evagination of the pharynx toward the stomodeum in amphibians; contact with the stomodeum forms the oral plate, which subsequently ruptures to open the mouth.

oral field figs. 54, 55

also see oral lobe

In starfish embryos, a depressed ventral area surrounding the stomodeum or mouth.

oral hood figs. 56, 71

In Amphioxus, a thin-walled funnel-like structure of the head leading to the mouth; fringed by stiff tentacles or cirri.

oral lobe figs. 39, 53–55

also see oral field

In starfish embryos, a projecting area above the oral field and separated from it by the preoral ciliary band.

oral plate figs. 163, 175, 187, 213

also see stomodeum

A double-layered membrane composed of the floor of the stomodeum and the anterior wall of the pharynx; rupture of the membrane opens the mouth into the pharynx. *Syn.:* pharyngeal membrane.

ostium of oviduct fig. 145

The opening of the oviduct into the peritoneal cavity near the ovary; accepts the egg into the oviduct after ovulation.

otic placode   figs. 97, 171, 178
also see otic vesicle
A thickening of head ectoderm lateral to
the myelencephalon; invaginates, forming
the otic pit, and separates from the
ectoderm as the otic vesicle; subsequently
forms the inner ear and contributes cells
to the ganglion of the 8th cranial nerve.
Syn.: auditory placode; ear placode.

otic vesicle   figs. 103, 111, 115, 117, 120,
124, 130, 135, 136, 183–185, 192, 203,
204, 209, 211, 224, 229, 231, 232,
234, 235, 239–244
also see otic placode
A closed chamber formed by the
invagination of the otic placode;
separates from the head ectoderm and
subsequently forms the inner ear.
Syn.: auditory vesicle; ear vesicle.

ovarian lumen   fig. 24
The cavity within the ovary of sea urchins
into which eggs are ovulated; leads to the
exterior through the genital pore.

ovarian wall   fig. 24
Connective tissue capsule enclosing the
germ cells and accessory cells in the ovary
of sea urchins.

ovary   figs. 7–9, 145
The female gonad; organ where mature
eggs form and ovulate; in vertebrates,
also secretes the female sex hormones
estradiol and progesterone.

pachytene stage   figs. 3, 6, 12–14
also see diplotene stage
A stage of the first maturation division in
spermatogenesis and oogenesis, follows
the zygotene stage during which synapsis
of homologous chromosomes occurs;
contains a haploid number of bivalents
(tetrads) or double chromosomes in
synapsis; bivalents shorten and thicken;
followed by the diplotene stage.

pancreas   figs. 120, 232
also see dorsal pancreas; ventral pancreas
A digestive and endocrine gland arising
as outgrowths of the duodenum and liver
diverticulum.

pancreatic duct   fig. 265
also see dorsal pancreas; pancreas;
ventral pancreas

The tubular connection between
pancreatic tissue and the duodenum;
formed by outgrowth of both the dorsal
and ventral pancreas; usually only the
ventral duct persists.

parathyroids   fig. 248
Masses of endocrine gland tissue usually
derived from the 3rd and 4th pharyngeal
pouches; migrate to the vicinity of the
thyroid.

pectinate muscles   fig. 254
Muscles forming ridges on the inner
surface of the atrial walls of the heart;
identify the primitive part of the atria.

pelvis of metanephros   figs. 235, 268
also see metanephric diverticulum
The expanded distal end of the
metanephric diverticulum; is surrounded
by metanephrogenic mesenchyme; forms
the pelvis, calyces, and collecting tubules
of the metanephros.

pericardial cavity   figs. 135, 215, 250,
251, 252, 255
also see pericardial coelom
The body cavity around the heart; derives
from the pericardial coelom as the latter
becomes isolated from the
pleuroperitoneal coelom.

pericardial coelom   figs. 103, 107, 111,
116, 124, 125, 168, 169, 171, 173, 177,
195–197
also see pericardial cavity
The large coelomic space around the
heart; formed by a cleft in the lateral
plate mesoderm of the head; part of the
splanchnic mesodermal layer thus formed
develops into heart; is cut off from the
pleuroperitoneal coelom by the
pleuropericardial membranes to from
the pericardial cavity.

peripheral cytoplasm   fig. 142
In the chicken oocyte, an outer or cortical
layer of finely granular cytoplasm; as the
oocyte matures it contributes to the
blastodisc.

peripheral syntrophoblast   figs. 273, 274
also see villus syntrophoblast; placental
villi; basal plate

An extension of the villus syntrophoblast
covering the fetal surface of the basal

p

plate; like villus syntrophoblast it is in contact with maternal blood and underlaid by cytotrophoblast.

peritoneal cavity figs. 126, 139, 216, 217, 261, 269
also see embryonic coelom
The body cavity of the abdomen; derives from the posterior region of the coelom; after the pericardial cavity is cut off from the pleuroperitoneal cavity, the latter is split into two pleural cavities and a peritoneal cavity by the pleuroperitoneal membranes.

perivitelline space figs. 18-21, 44-46, 71
The space between the fertilization membrane and the egg surface; between the zona pellucida and the egg in mammals; contains perivitelline fluid, the "culture medium" of the developing egg.

petrosal ganglion figs. 235, 245
also see ganglion of cranial nerve 9
The more ventral of two ganglia of the 9th cranial nerve; contributes sensory fibers to the nerve. *Syn.:* inferior ganglion.

pharyngeal membrane *see* oral plate

pharyngeal pouch fig. 125
also see pharyngeal pouches 1; 2; 3; 4
Paired evaginations of the lateral pharyngeal wall; meet corresponding ectodermal invaginations (branchial grooves) to form the branchial clefts; lie between the branchial arches.

pharyngeal pouch 1 figs. 98, 102, 106, 107, 110, 118, 119, 192, 212, 245
also see branchial cleft 1
Paired evaginations of the lateral pharyngeal wall posterior to the mandibular arch; form the endodermal part of the first branchial clefts; extend dorsally toward the otic vesicles to form the Eustachian tubes and tympanic cavities. *Syn.:* branchial pouch 1; gill pouch 1; visceral pouch 1.

pharyngeal pouch 2 figs. 102, 118, 212, 213, 235, 246
also see branchial cleft 2
Paired evaginations of the pharyngeal wall posterior to the hyoid arch; form the endodermal part of the second branchial

clefts; subsequently obliterates, except contributes to the thymus in lower vertebrates (frog). *Syn.:* branchial pouch 2; gill pouch 2; visceral pouch 2.

pharyngeal pouch 3 figs. 102, 106, 111, 118, 195, 212, 213, 229, 235, 246, 248
also see branchial cleft 3
Paired evaginations of the pharyngeal wall posterior to the third branchial arch; form the endodermal part of the third branchial clefts; contributes to the thymus and parathyroids. *Syn.:* branchial pouch 3; gill pouch 3; visceral pouch 3.

pharyngeal pouch 4 figs. 106, 213, 214, 249
also see branchial cleft 4
Paired evaginations of the lateral pharyngeal wall posterior to the fourth branchial arch; contribute to the parathyroids.

pharynx figs. 56, 69-71, 73, 97, 101-104, 106, 107, 110, 111, 115-118, 124, 130, 131, 133-136, 173, 175-177, 183-185, 187, 192-196, 207, 208, 214, 226, 228, 232, 234, 237, 238, 248, 250, 251
also see foregut
The region of the embryonic foregut bearing branchial clefts; large in Amphioxus, fishes and amphibian tadpoles and forms gills; in air breathers extends posteriorly to the glottis; is much reduced and its pouches are transformed.

pia mater fig. 242
The inner layer of the meninges; a delicate membrane on the brain and spinal cord derived from head mesenchyme and neural crest.

pigmented cortex figs. 83, 84
The surface coat of amphibian eggs and early embryos; a gelled layer containing much melanin in the animal hemisphere.

pigmented layer of optic cup figs. 122, 134, 215, 246
also see optic cup
The outer wall of the optic cup; formed from the medial half of the optic vesicle; forms the pigmented epithelium of the retina, ciliary body, and iris.

pigment spot fig. 73
Pigment cells associated with the neural tube in Amphioxus; light-sensitive organs.

pituitary   *see* hypophysis

placental barrier   fig. 274
also see placental villi
The layers of the placenta interposed between maternal blood and fetal blood; includes in humans at least the chorionic endothelium of fetal capillaries and the syntrophoblast but in some areas also the cytotrophoblast and chorionic mesenchyme; controls the exchange between maternal the fetal bloods.

placental villi   figs. 273, 274
also see anchoring villus; stem placental villus; peripheral syntrophoblast; villus cytotrophoblast; fetal capillaries
A branching tree-like outgrowth of the chorion into the maternal blood of the intervillus spaces; covered by two epithelial layers, the outer syntrophoblast and inner cytotrophoblast; contains a mesenchyme connective tissue and fetal blood vessels; some villi attach to the maternal decidua; forms the placental barrier which interposes between the maternal and fetal bloods, controlling the exchange of substances between the two blood streams.

placode derived ganglionic cells
fig. 191
also see ganglion of cranial nerve 5
Those ganglionic neuroblasts derived from ectodermal thickenings, the placodes; the other neuroblasts arise from neural crest.

pleural cavities   figs. 214, 255, 256
The body cavities surrounding the lungs; derive from anteriodorsal divisions of the pleuroperitoneal coelom, which become isolated from the pericardial and peritoneal cavities.

pluteus larva   figs. 38–40
A bilateral, free-swimming larval stage of sea urchins, sand dollars, and brittle stars; possesses long ciliated arms; after a growth period, larvae fall to the bottom and metamorphose into adults.

polar body I   figs. 18, 21, 22, 44, 57
A small cell separated from the primary oocyte by the first maturation division; may divide again but then degenerates.

polar body II   fig. 19, 23, 57, 74
A small cell separated from the secondary oocyte by the second maturation division; degenerates.

postanal gut   fig. 239
The extension of the hindgut into the tail; gradually degenerates. *Syn.:* tail gut.

postcardinal veins   *see* posterior cardinal veins

posterior cardinal veins   figs. 118, 120, 127, 172, 185–187, 199, 200, 206, 207, 209, 215–220, 225, 226, 227, 231, 232, 236, 238, 256, 257, 261–263, 265, 266
The primitive paired veins of the trunk; lie dorsal to the mesonephros and drain with the anterior cardinals into the common cardinals; as the mesonephroi grow, form the renal portal veins; mostly degenerate in amniotes with the mesonephros, but form the iliac veins. *Syn.:* postcardinal veins.

posterior intestinal portal   figs. 185, 202, 203, 221
also see midgut
The opening from the midgut into the hindgut of amniotes; moves anteriorly, lengthening the hindgut; meets the anterior intestinal portal at the level of the small intestine to form the yolk stalk.

posterior liver diverticulum   figs. 198, 217
also see anterior liver diverticulum; liver
One of two outgrowths of the duodenum of birds which grow, branch, and anastomose to form the liver. *Syn.:* caudal liver bud; ventral liver bud.

posterior vena cava   figs. 225, 236–238, 256, 257, 259, 260–263, 265
The principal systemic vein of the trunk; derives from several primitive paired veins, including the right vitelline, right subcardinal, right supracardinal, and right posterior cardinal; originally drains into the sinus venosus but is carried into the right atrium as the sinus venosus merges with the atrium. *Syn.:* inferior vena cava, post cava.

posterior vitelline vein   fig. 186
also see vitelline veins
A branch of the left vitelline vein extending posteriorly to receive the sinus terminalis.

prechordal plate   figs. 90, 91, 93, 101, 163
A mass of mesodermal cells anterior to the notochord and between the foregut and

p

prosencephalon; constitutes, with the notochord, the axial mesoderm; a site of head mesenchyme formation.

preoral gut   figs. 207, 208, 213, 226, 229
The projecting tip of the foregut anterior to the oral plate; gradually atrophies. *Syn.*: Seesell's pocket.

preoral pit   fig. 73
also see left diverticulum
The external opening into the left diverticulum in Amphioxus; located anterior to the mouth; arises as an invagination of ectoderm.

presumptive fate map   figs. 74, 91, 92, 152
A graphic representation of the location of organ-forming regions drawn on an earlier embryonic stage — often on the late blastula or early gastrula stages. *Syn.*: prospective fate map.

primary bronchi   *see* lung buds

primary follicle   figs. 7–9
A small follicle of the mammalian ovary with but one layer of follicle cells surrounding an oocyte; the smallest or primordial follicles were formed during fetal life; they grow in response to follicle-stimulating hormone of the hypophysis.

primary mesenchyme   figs. 34–37
A loose cluster of cells near the vegetal pole and within the blastocoel of sea urchins; derives from micromeres; contributes to the skeleton of the pluteus.

primary oocyte   figs. 7, 8, 16, 24, 27, 144
also see oocyte
The immature egg prior to completion of the 1st maturation division.

primary spermatocytes   figs. 1, 2, 3, 5, 12–14, 77, 78, 142, 143
Large germ cells of the testis formed by growth of spermatogonia; undergo the first maturation division to form secondary spermatocytes.

primitive folds   figs. 154, 157, 158
also see primitive streak
The thickened ridges of the primitive streak; formed by convergent flow of epiblast.

primitive groove   figs. 154, 157, 158
also see primitive streak

A depressed trough between the primitive folds; a region of involution of epiblast cells into the mesoderm and endoderm.

primitive knot   figs. 152, 154, 156, 159, 160
also see primitive streak
The anterior thickened end of the primitive streak. *Syn.*: Hensen's node.

primitive streak   figs. 152, 154, 159, 160, 170, 171, 181
A longitudinal thickening in the epiblast of early amniote embryos; formed by convergent flow of epiblast toward the caudal midline; site of involution of epiblast cells into the mesoderm and endoderm; consists of parallel longitudinal ridges (primitive folds), separated by a primitive groove, and an anterior thickening, the primitive knot.

primordial follicles   figs. 7–9
also see primary follicle
Small ovarian follicles just within the tunica albuginea; consist of a small oocyte and a thin layer of follicle cells; formed during fetal life; become primary follicles as their growth begins.

proamnion   figs. 154, 159, 163, 174
A crescent-shaped area lacking mesoderm around the head of early bird embryos; initially delimits the anterior end of the embryo; later drawn under the head by the head fold, invaded by the mesoderm, and contributes to the amnion.

prochromosome stage   fig. 12
also see lepototene stage
The earliest prophase stage of the first maturation division of spermatogenesis of some insects; chromosomes contract into discrete bodies of which there are a diploid number; unraveling of the prochromosomes leads to the next or leptotene stage.

proctodeum   figs. 101, 103, 104, 114, 129
An ectodermal invagination on the ventral side of the trunk at the base of the tail; breaks into the hindgut to form the anus. *Syn.*: anal pit.

pronephric duct   figs. 120, 127, 128, 200
also see mesonephric duct; pronephros
A tubule connecting the pronephros with the cloaca; arises from caudal growth of

pronephric buds; subsequently becomes
the mesonephric duct.

pronephric tubule   figs. 117, 200, 230
also see nephrotome; pronephric duct;
pronephros
Tubules derived from the anterior
nephrotomes, one pair per segment; bear
a nephrostome opening into the coelom on
the distal end; proximal end connects to
pronephric duct; degenerates before adult
stage but functional during larval period in
amphibians.

pronephros   figs. 113, 120, 126, 130,
137–139, 184, 218
also see nephrotome, pronephric duct,
pronephric tubules. The first and most
anterior kidney to form; derives from
buds of nephrotomes which hollow
out to form tubules—one pair per body
segment; one end of each tubule opens
as a nephrostome into coelom; the tubules
link together to form the pronephric duct
which grows posteriorly along the somites
to the cloaca; is vestigial in amniotes but
large in lower vertebrates and functions in
the larval stage; the pronephric duct is
appropriated by the mesonephros in
amniotes.

pronucleus   figs. 19, 26, 57
A haploid nucleus found in fertilized eggs,
one pronucleus derives from the sperm
and a second one from the egg; may fuse
or enter prophase of the 1st cleavage
division separately.

prosencephalon   figs. 96, 97, 101–109,
115, 116, 120, 170–174, 185
also see diencephalon; telencephalon
The anterior primary brain vesicle; forms
two lateral evaginations (optic vesicles)
and a ventral evagination (the
infundibulum); then differentiates
into an anterior telencephalon and a
posterior diencephalon.
Syn.: forebrain.

pulmonary aorta   fig. 251
The trunk of the pulmonary arteries;
connects with the right ventricle; derives
from the bulbus cordis by longitudinal
division of the latter.

pulmonary arteries   figs. 225, 235, 236, 253
Connect the pulmonary aorta with the

lungs; basal sections derive from the
6th aortic arches.

pulmonary vein   fig. 255
The vessel carrying blood from the lungs
to the left atrium; arises as an outgrowth
of the left atrium or in birds from sinus
venosus and connects with the pulmonary
plexus.

rami n. 5   fig. 235
also see mandibular ramus; cranial n. 5;
maxillary ramus; cranial n. 5; ophthalmic
ramus; cranial n. 5
The main branches of the fifth cranial
nerve (trigeminal n.) consisting of, from
anterior to posterior: ophthalmic ramus,
maxillary ramus, mandibular ramus.

Rathke's pouch   figs. 184, 185, 187, 192,
204, 207, 214, 226, 229, 237, 246
also see hypophysis
A dorsal evagination of the stomodeum
extending under the diencephalon to the
infundibulum in amniotes; becomes
isolated from the stomodeum and forms
the pars distalis (anterior lobe), the pars
intermedia (intermediate lobe), and pars
tuberalis of the hypophysis.

rectum   figs. 145, 266, 267
The posterior segment of the large
intestine; formed by splitting off from
the dorsal side of the cloaca.

reduction division   *see* maturation division

residual bodies   fig. 142
Granules of degenerating cytoplasm
sloughed off differentiating spermatids;
phagocitized by the Sertoli cells.

residual spermatogonium   fig. 77
Large reserve germ cells of amphibia; may
proliferate mitotically to replace cells which
have matured into sperm; each is enclosed
by a follicle cell.

retina   figs. 130, 134, 215, 246
also see nervous layer of optic cup; optic
cup; pigmented layer of optic cup
The inner sensory and pigmented layers of
the eye; derived from the optic cup whose
inner wall becomes the sensory-nervous
part of the retina and an outer wall
forming the pigmented epithelium of the
retina.

rhombencephalon   figs. 97, 103, 104,

110–112, 115, 116, 123–125, 173, 176–178
also see metencephalon; myelencephalon
The third and posterior primary brain vesicle extending from the mesencephalon to the spinal cord; divides into an anterior metencephalon and a posterior myelencephalon. *Syn.*: hindbrain.

right atrium   figs. 235, 237, 251–254
also see atrium
The right division of the primitive atrium separated from the left atrium by the atrial septum; receives blood from the sinus venosus (or later from the vena cavae) and delivers it through the interatrial foramen to the left atrium and through the right atrioventricular canal to the right ventricle; after breathing begins, the interatrial foramen closes.
Syn.: right auricle.

right auricle   *see* right atrium

right diverticulum   fig. 73
also see dorsal diverticulum
In Amphioxus, arises from the dorsal diverticulum, extends ventrally, and expands to form the thin-walled head cavity.

right horn of sinus venosus   fig. 253
also see sinus venosus
The part receiving blood from the right common cardinal, right vitelline, right umbilical veins, and, later, posterior vena cava; eventually incorporated into the right atrium with its veins.

right umbilical vein * *see* umbilical vein

right ventricle   figs. 235, 252, 253, 255, 256
also see ventricle
A thick-walled heart chamber formed from the partitioning of the primitive ventricle by the ventricular septum; receives blood from the right atrium and delivers it to the pulmonary aorta and ductus arteriosus.

right vitelline artery   fig. 186
also see vitelline arteries
The arterial supply for the right half of the yolk sac.

right vitelline vein   fig. 186
also see vitelline veins
The venous return for the right half of the yolk sac.

root of cranial nerve 5   figs. 211, 242
also see cranial nerve 5
The part of the fifth cranial nerve connecting the semilunar ganglion to the metencephalon.

sclerotome   figs. 126, 199, 215, 246, 251, 262, 270, 271
also see somites
The medial, mesenchymatous division of the somite; arises from cells of the medioventral wall of the early somite; envelops the notochord and spinal cord; sclerotomes split transversely and adjacent halves fuse to form the rudiments of the vertebrae and ribs.

secondary mesenchyme   figs. 35–37
also see primary mesenchyme
Cells which migrate into the blastocoel from the wall of the archenteron during gastrulation in sea urchins; occupies the animal part of the blastocoel forming skeleton and muscle.

secondary spermatocytes   figs. 3, 10, 13, 143
Male germ cells formed from primary spermatocytes by the 1st maturation division; undergo at once the 2nd maturation division to form spermatids; distinguished from both primary spermatocytes and spermatids by intermediate size.

Seesell's pocket   *see* preoral gut

segmental mesoderm   figs. 159, 167, 170, 171
Paraxial mesoderm extending posteriorly from the last somite; will form somites by segmentation.

segmentation cavity   *see* blastocoel

semilunar ganglion   figs. 234, 235, 242–245
also see ganglion of cranial nerve 5.

seminiferous tubules   figs. 1, 76, 141
Tubules within the testis; bounded by a thin basement membrane of connective tissue and containing the male germ cells and Sertoli cells.

sensory root   *see* dorsal root of spinal nerve

septum   fig. 76
In the frog testis, connective tissue

membranes enclosing the seminiferous tubules.

septum spurium   fig. 254

A temporary ridge on the dorsal wall of the right atrium extending to the valves of the sinus venous (valvulae venosae).

serosa   *see* chorion

Sertoli cell   figs. 2, 5, 77, 142

The sperm nurse cell of vertebrates; in mammals, a tall, columnar, phagocytic cell extending from the basement membrane to the lumen of the seminiferous tubule; the outline of the cell is irregular and obscure; the nucleus is light-staining with a prominent nucleolus; differentiating germ cells become embedded in cytoplasmic pockets in Sertoli cells and withdraw at maturity; forms part of the blood–testis barrier. *Syn.*: sustentacular cells.

shell   figs. 146, 188, 227

also see egg shell

The outer wall of the egg; in birds composed of calcium carbonate crystals impregnating protein fibers; perforated by numerous microscopic pores to permit respiration; the pores are sealed by the cuticle composed of dry albumen; reinforced by the shell membranes attached to its inner surface; the shell is secreted around the egg while it lies in the uterus.

shell membranes   figs. 146, 188, 229

In the bird egg, a pair of flexible membranes composed of protein fibers and attached to the shell which they reinforce; at the blunt end of the egg the membranes separate to enclose the air chamber which they immobilize.

sinus venosus   figs. 107, 119, 120, 136, 183-187, 197, 203, 204, 206-209, 215, 224-226, 228, 237-239

also see left horn of sinus venosus; right horn of sinus venosus; transverse sinus venosus

Initially the most posterior chamber of the heart, receiving the venous return and delivering it to the atrium; after the partitioning of the atrium, empties into the right atrium; disappears as a heart chamber by atrophy and by incorporation into the atria; originates the heart beat and later transfers that function to the atrium by forming the sinoatrial node.

skeleton   fig. 42

In sea urchin embryos and larvae, calcite spicules formed by mesenchyme cells; support the arms of the pluteus larva.

small intestine   figs. 262, 263, 265-267

also see duodenum; intestinal loop; intestine

The segment of gut following the stomach; arises from foregut and midgut in amphibians, from foregut and hindgut in amniotes.

smooth muscle   fig. 272

also see myometrium

Nonstriated involuntary muscle located in the walls of hallow organs as in the uterus; contraction controlled by hormones (oxytocin, adrenalin, noradrenalin) and the autonomic nervous system.

somatic mesoderm   figs. 113, 164, 173, 182, 219

also see lateral plate mesoderm; somatopleure

The cellular layer immediately outside the coelom; arises by splitting from lateral plate mesoderm; forms parietal peritoneum and, by fusion with myotomes, dermatomes and epidermis, forms body wall and limbs; in the extraembryonic area, fuses with ectoderm to form the somatopleure of the amnion and chorion.

somatopleure   figs. 164, 169, 175-178, 182, 185, 188

also see amnion; chorion

A double-layered membrane composed of ectoderm and somatic mesoderm; contributes to the body wall and extends into the extraembryonic area; forms the amniotic folds which by enveloping the embryo transform extraembryonic somatopleure into amnion and chorion.

somites   figs. 68, 73, 91, 95, 99, 103-105, 112-118, 125, 127-129, 139, 140, 166, 170, 171, 173, 179, 183, 185, 187, 195, 198, 201, 203, 204, 207-209, 211, 212, 215, 220, 224, 226, 227, 231, 232-234, 238, 239, 243, 261

also see dermatome; myotome; sclerotome

The segments of paraxial mesoderm; form first at the posterior end of the myelencephalon and extend progressively

as a series of paired blocks posteriorly
into the tail; are separated by intersomitic
grooves and attach laterally to the
nephrotomes; are primary segments of the
body which establish all other segmental
patterns; differenatiate into a lateral
dermatome, a middle myotome, and
medial sclerotome. *Syn.:* epimere.

sperm   figs. 1, 2, 3, 5, 10, 16–18, 76–78
also see differentiating spermatid;
spermiogenesis
The mature male germ cell; in vertebrates,
a small, haploid, highly specialized,
flagellated cell which can attach to and
penetrate egg membranes to activate the
egg; formed from a spermatid through a
complex differentiation called
spermiogenesis. *Syn.:* spermatozoon.

spermatid   figs. 1, 2, 5, 10, 13–15, 76, 77,
142
also see differentiating spermatid
A small haploid germ cell of the testis;
formed from a secondary spermatocyte by
the 2nd maturation division; embeds in a
pocket within the Sertoli cell and
differentiates into a sperm.

spermatocyte   figs. 6, 76
also see primary spermatocytes; secondary
spermatocytes

spermatogonia   figs. 1, 2, 5, 6, 10, 142
The "stem" germ cells of the testis;
divide mitotically to regenerate the
germinal epithelium against the loss of
mature sperm; located near the basement
membrane of the seminiferous tubule
outside the blood-testis barrier; may
enter a prolonged growth phase forming
primary spermatocytes.

sperm heads   figs. 2, 77
also see sperm; sperm tails
That part of the sperm containing the
nucleus and acrosome.

spermiogenesis   figs. 78, 143
also see differentiating spermatid
The final phase of spermatogenesis during
which the spermatid transforms or
differentiates into a sperm; during this
period the cells are enveloped by
cytoplasm of the Sertoli cells which
probably provide the special environment
required for the transformation.

sperm tails   figs. 1, 141, 142
also see sperm; differentiating spermatid;
immature sperm; centrioles
The flagellum of the sperm; derived from
the cytoplasm of the spermatid; composed
of an axial filament arising from centrioles
near the head, a mitochondrial sheath,
fibers, and a plasma membrane.

spinal cord   figs. 99, 101, 103–105, 113–117,
121, 126–131, 139, 140, 170, 179, 183, 187,
197, 201–204, 212, 213, 223, 226, 227, 228,
232, 237, 238, 248–251, 255, 263, 269–271
also see neural tube
The central nervous system posterior to
the brain; derives from posterior neural
tube; bears a pair of spinal nerves for
each body segment; wall differenatiates
into an inner ependymal layer, a middle
mantle layer, and an outer marginal layer;
the latter two layers are rudiments of
gray matter and white matter, respectively.

spinal ganglia   figs. 120, 127, 129, 140,
197, 212, 213, 216, 232, 235, 237, 239,
246, 248, 249, 251, 255, 259, 261, 270,
271
Ganglia borne on dorsal roots of spinal
nerves; derive from neural crest and supply
sensory nerve fiber of the spinal nerve.
Syn.: dorsal root ganglia.

spinal nerves   fig. 270
also see dorsal root of spinal nerve;
spinal ganglia; ventral root of spinal nerve
Paired nerves emerging from spinal cord at
each body segment; each is connected to
the spinal cord by a dorsal root bearing a
spinal ganglion and by a ventral root; the
spinal nerve trunk divides immediately into
a dorsal ramus and a ventral ramus; a
ramus communicans connects to
autonomic ganglia

spinal nerve trunk   fig. 261
also see dorsal root spinal nerve; spinal
nerves; ventral root spinal nerve; ventral
ramus spinal nerve
Formed by the junction of the dorsal and
ventral roots; branches immediately into
the dorsal ramus, ventral ramus and ramus
communicans.

splanchnic mesoderm   figs. 113, 164,
169, 173, 182, 219
also see lateral plate mesoderm;
splanchnopleure

The cell layer between the coelom and endoderm; arises by splitting from lateral plate mesoderm; fuses with endoderm to form the wall of the gut and respiratory tract; forms mesenteries, visceral peritoneum, heart, and germinal epithelium; in the extraembryonic area, fuses with endoderm to form the splanchnopleure of the yolk sac and allantois.

splanchnopleure   figs. 169, 172, 175–178, 182, 185, 188, 210
also see allantois; yolk sac
A double membrane composed of splanchnic mesoderm and endoderm; forms the gut wall and extends into the extraembryonic area to form the yolk sac and allantois.

stem bronchus   fig. 257
The most posterior secondary bronchus of the developing lung; forms the lower lobe of the lung.

stem placental villus   fig. 273
also see placental villi
In the placenta, the trunk or large branch of the placental villus; contains arteries and veins with fetal blood which arise from the umbilical arteries and veins; supports terminal villi.

stomach   figs. 38–42, 54–56, 120, 130, 137, 138, 197, 216, 232, 235, 239, 259, 261, 262, 264
An enlarged segment of the foregut posterior to the esophagus; derives from archenteron in the sea urchin and starfish and Amphioxus; the lining epithelium and glands form from gut endoderm but the muscle, blood vessels, and connective tissue develop from splanchnic mesoderm; in birds, the stomach differentiates into a proventriculus and a gizzard.

stomodeum   figs. 37–40, 43–54, 76–96, 101, 103, 104, 107, 109, 115, 116, 119, 120, 122, 173, 175, 183–185, 187, 193, 203, 204–210, 213, 214, 230, 232
also see mouth
The ectodermal rudiment of the mouth; an invagination in the anterioventral ectoderm of the head which contacts the anterior wall of the foregut; its floor is the oral plate; rupture of the membrane opens the mouth into the pharynx; a rudiment of the hypophysis, called Rathke's pouch in amniotes evaginates from the dorsal wall of the stomodeum.

stratum granulosa   figs. 7, 8, 144
also see cumulus oophorus
The inner stratified epithelium of large ovarian follicles; derives from follicle cells of primary follicles; in mammals, contributes to the corpus luteum after ovulation.

stroma   fig. 272
also see connective tissue; endometrium
The connective tissue framework of an organ; in the mammalian ovary and endometrium consists of a dense population of elongate cells and some delicate fibers

subcardinal anastomosis   fig. 269
also see subcardinal veins
A medial interconnection between the right and left subcardinal veins; contributes to the prerenal segment of the posterior vena cava.
Syn.: subcardinal sinus.

subcardinal veins   figs. 225, 237, 266
Primitive paired veins of the trunk; they lie ventral to the mesonephroi and parallel to the posterior cardinals which they mostly replace; they subsequently contribute to the posterior vena cava and its branches.

subchorda   *see* subnotochordal rod

subcephalic pocket   figs. 161, 163, 174
also see head fold
A cavity beneath the embryonic head formed by the head fold as it pushes under the head; lined by ventral head ectoderm and ectoderm of somatopleur; lengthens as the head grows forward and the head fold is drawn posteriorly.

subclavian artery   figs. 225, 235, 236
The artery of the shoulder and forelimb; arises by the enlargement of the 7th intersegmental artery; in mammals, the right subclavian also receives contributions from the right 4th aortic arch and right dorsal aorta.

subclavian vein   figs. 225, 236, 253
The vein of the forelimb; connects at first

to posterior cardinal but later shifts to the anterior cardinal.

subgerminal cavity   figs. 149, 150, 155, 163

A space beneath the hypoblast of the area pellucida in birds; becomes the midgut cavity; communicates through the intestinal portals with the foregut and hindgut.

subintestinal vein   fig. 225

A vein in pig embryos extending from the base of the tail along the ventral margin of the intestine to the vitelline veins; initially drains the allantois, posterior limb buds, and intestine; mostly replaced by the development of the allantoic veins.

subnotochordal rod   figs. 104, 114, 117

A strand of cells lying between the midgut and notochord in amphibians; of endodermal origin; degenerates. *Syn.*: hypochordal rod; subchorda.

superior ganglion   figs. 235, 242

also see ganglion of cranial nerve 9

The dorsal ganglion of the 9th cranial nerve; with the petrosal ganglion, supplies sensory fibers to the nerve.

superior mesenteric artery   figs. 263, 265, 266–268

also see vitelline arteries

The arterial supply of the small intestine; derived from the vitelline arteries

superior mesenteric vein   fig. 266

The main branch of the hepatic portal vein; drains the digestive tract.

sustentacular cells   see Sertoli cell

sympathetic ganglia   figs. 253, 261

A series of paired ganglia dorsal to the aorta and connected to the spinal nerves by the rami communicans; part of the autonomic nervous system; derives from neural crest.

syntrophoblast   see villus syntrophoblast

systemic arch   see aortic arch 4

tail   fig. 262

also see tail bud

The extension of the body posterior to the anus; derives from the tail bud.

tail bud   figs. 101, 102, 104, 183, 187, 193, 203, 207, 208, 224, 230, 232

The rudiment of the tail and posterior trunk; a mass of undifferentiated tissue projecting from the posterior end of the embryo; derives from primitive streak in amniotes; contributes to neural tube, somites, and notochord.

tail fin   figs. 105, 106, 116–119, 130, 140

also see caudal fin

A blade-like extension of the border of the tail in amphibians and Amphioxus; continuous anteriorly with the dorsal fin.

tail fold   figs. 183, 202, 222

A depressed fold encircling the tail bud and connecting anteriorly with the body folds; forms part of the boundary between the embryonic and extraembryonic areas; undercuts the tail bud and posterior trunk forming hindgut.

tail gut   see postanal gut

telencephalon   figs. 121, 131, 133, 183, 184, 187, 195, 196, 203, 204, 207–210, 216, 217, 226, 227, 230, 232, 239, 249–251

also see cerebral hemispheres

The anterior division of the prosencephalon; the greatly enlarged roof forms the cerebral hemispheres; the floor forms the olfactory bulbs, hippocampus, and corpus striatum; the cavities are the lateral ventricles of the brain.

testicular cyst   fig. 10

also see testicular lobe wall

In the grasshopper testis, a compartment within a testicular lobe bounded by connective tissue septa and containing a group of germ cells at the same stage of maturation.

testicular lobe wall   figs. 10, 15

also see testicular cyst

In the grasshopper testis, the connective tissue capsule enclosing a lobe or major division of the testis; the lobe is divided into cysts, each containing a cluster of germ cells; the apical end of the lobe contains proliferating spermatogonia with more mature germ cells extending toward the opposite end which opens into a vas deferens.

testis   figs. 1, 2

The male gonad; the organ in which sperm

differentiate; secretes the male sex
hormone testosterone in vertebrates.

tetrad   *see* bivalent

theca externa   figs. 8, 79
The outer connective tissue layer of
Graafian follicles; arises from the stroma
of the ovary; is the outer wall of the ovary
in amphibians.

theca folliculi   figs. 7, 8
also see theca externa; theca interna
The outer capsule of ovarian follicles;
forms from connective tissue stroma as
follicles grow; in mature follicles
differentiates into theca interna and
theca externa.

theca interna   figs. 8, 79
A vascular layer between the theca externa
and the stratum granulosa of large ovarian
follicles; contains endocrine gland cells,
connective tissue, and blood vessels;
contributes to the corpus luteum after
ovulation or to the interstitial tissue after
follicular atresia in mammals.

third ventricle   figs. 187, 208, 237
also see lateral ventricles

Originally the enlarged neural canal of the
telecephalon; later divides into the lateral
ventricles of the cerebral hemispheres and
the definitive third ventricle of the
diencephalon; the thin roof forms a
choriod plexus.

thyroid   figs. 104, 110, 120, 123, 130,
131, 134, 195, 207, 208, 213, 226, 232,
248

An endocrine gland in the throat region;
forms as a ventral diverticulum of the
pharynx at the level of the 2nd branchial
arch; the rudiment bifurcates and migrates
posteriorly, becoming isolated from the
pharynx.

tooth   figs. 131–133
In tadpoles 3 or 4 rows of horny epidermal
papillae attached to the jaw cartilages;
frequently shed and replaced; lost during
metamorphosis and replaced by true teeth.

tongue   figs. 234, 237–239
also see copula; lateral swellings;
tuberculum impar
In mammals, arises by fusion of several
elevations on the floor of the mouth and

pharynx; these elevations include two
lateral swellings and a median tuberculum
impar on the mandible, the copula on the
hyoid arch, and contributions from the
3rd and 4th branchial archs; later the
embryonic tongue is invaded by muscle
from a more posterior level and its
innervation by the 12th cranial nerve.

trabeculae carneae   fig. 254
Interlacing muscle bands in the wall of the
ventricle of the heart; contribute to the
formation of the ventricular septum.

trachea   figs. 131, 226, 235, 252, 253
The part of the respiratory tract
connecting the laryngotracheal groove
with the lung buds or, later, the larynx
with the primary bronchi; arises with the
lung buds as a ventrocaudal diverticulum
of the pharynx; muscle and connective
tissues develop from the splanchnic
mesoderm of the ventral mesoesophagus.

transverse septum   figs. 120, 239, 256, 257
A mass of mesenchyme posterior to the
heart, incompletely separating the
pericardial cavity from the peritoneal
cavity; encloses the veins that enter the
heart; the liver is attached to its caudal
face; contributes to the diaphragm in
mammals.

transverse sinus venosus   fig. 256
also see sinus venosus

A narrow middle part carrying blood from
the left horn to the sinoatrial opening;
eventually forms the coronary sinus.

trigeminal nerve   *see* cranial nerve 5

trigeminal placode   fig. 191
also see ganglion of cranial nerve 5;
neural crest

A thickened plate of head ectoderm dorsal
to the mandibular arch; cells detach from
the under surface and join neural crest
cells forming the semilunar ganglion of the
fifth cranial nerve.

trophoblast   fig. 275
also see villus cytotrophoblast; villus
syntrophoblast

The outer epithelial covering of the
chorion and placental villi, the principle
component of the placental barrier;
probably secretes the placental hormones.

tuv

truncus arteriosus *see* ventral aorta

trunk organizer fig. 91

Inductor of trunk parts; consists of middle and posterior notochord and the somites; in amphibian embryos, follows the head organizer over the dorsal lip during gastrulation. *Syn.*: trunk inductor.

tuberculum impar fig. 246

also see tongue

A median elevation on the mandible in the floor of the mouth; fuses with the lateral swellings to form the body of the tongue.

tuberculum posterius fig. 104

An elevation on the floor of the diencephalon marking its posterior boundary.

tunica albuginea figs. 1, 7, 8, 77

A fibrous connective tissue capsule or membrane enveloping the ovary and testis.

ultimobranchial body fig. 250

An evagination from the caudal surface of the fourth pharyngeal pouch; may represent the fifth pouch; fuses with the thyroid rudiment probably forming the parafollicular cells of the thyroid gland. *Syn.*: postbranchial body.

umbilical arteries figs. 225, 235–238, 262, 263, 265–268

The arterial blood supply to the chorioallantois of birds and the placenta of mammals; a pair of vessels arising from the posterior end of the aorta; forms the common iliac and hypogastric arteries in mammals, and, after birth, the lateral umbilical ligaments. *Syn.*: allantoic arteries.

umbilical cord figs. 233, 234, 239, 262, 263

The narrowed connection of the embryo to the extraembryonic membranes; the outer wall is amnion and may contain the yolk stalk, allantoic stalk, vitelline blood vessels, umbilical blood vessels, and a gelatinous connective tissue; in birds, separates from the umbilicus just before hatching; in mammals, is bitten in two after birth; the strump dropping off in a few days.

umbilical vein figs. 176, 219, 225, 228, 231, 235, 239, 260, 262, 263, 265–268

Initially, paired embryonic vessels draining the allantois of birds or the placenta of mammals; the left vein atrophies early, the right atrophies after hatching or birth, forming, in mammals, the ligamentum teres. *Syn.*: allantoic vein.

ureter figs. 235, 268

The excretory duct of the metanephros; derives from the stalk of the metanephric diverticulum connecting at first with the mesonephric duct, its site of origin; later shifts to the cloaca in birds or to the urinary bladder in mammals. *Syn.*: metanephric duct.

ureteric bud *see* metanephric diverticulum

urogenital sinus figs. 234, 235, 238, 267

also see cloaca

A chamber split from the ventral part of the cloaca of mammals; receives the mesonephric ducts, Mullerian ducts, and allantoic stalk; contributes to the bladder; forms the urethra and, in females, the vestibule of the vagina as well.

uterine cavity fig. 272

also see endometrium; uterine epithelium; uterine glands

The lumen of the uterus; lined by the uterine epithelium of the endometrium.

uterine epithelium fig. 272

also see endometrium; uterine glands

A simple columnar glandular epithelium with some ciliated cells lining the uterine cavity; part of the endometrium.

uterine glands figs. 272, 275

also see endometrium

Tubular glands of the endometrium; active during the secretory phase of the menstrual cycle releasing a nutrient mixture of glycogen, mucinogen, and fat which provides a supportive medium for the developing embryo prior to implantation.

uterus fig. 145

In birds, the large terminal segment of the oviduct; forms the egg shell as the egg is held there; passes the finished egg into the cloaca at laying; in viviparous mammals, reptiles, and some fishes, provides metabolic support for the developing young.

vagus nerve *see* cranial nerve 10

valve of sinus venosus   fig. 253
Valve of the sinoatrial opening.
Syn.: valvulae venosae.

valvulae venosae   *see* valve of sinus venosus

vegetal hemisphere   figs. 32, 66
also see animal hemisphere; vegetal pole
That half of the egg (oocyte) or early embryo containing the most yolk, the other half being the animal hemisphere; the vegetal pole lies at its center and opposite the animal pole.

vegetal pole   figs. 49, 50, 65, 85
also see vegetal hemisphere
The end of the embryonic axis centered in the yolky region of the egg; opposite the animal pole.

vein   fig. 141
In the testis, veins, arteries, and capillaries branch in the interstitial connective tissue between the seminiferous tubules; blood vessels do not penetrate the tubules.

velar plate   fig. 135
Laterally projecting plates on the floor of the pharynx of tadpoles which reduce the openings of the pharynx into the opercular chamber to narrow slits.

velum transversum   fig. 207
A transverse fold and groove in the roof of the prosencephalon marking the boundary between the telencephalon and diencephalon.

vena capitis   figs. 225, 236
The principal embryonic vein draining the venous plexi of the brain and later the dural sinuses; derived from the anterior segment of the anterior cardinal vein.

ventral aorta   figs. 168, 172, 175, 185, 186, 194, 206, 208, 214, 226, 232, 248, 249

The outlet of the embryonic heart; lies in the floor of the pharynx and conducts blood from the bulbus cordis to the aortic arches; forms the innominate arteries and the ascending aorta. *Syn.*: aortic sac; truncus arteriosus.

ventral ectoderm   figs. 68, 87
also see ectoderm; epidermis
The outer or ectodermal layer covering the ventral surface of the embryo; forms epidermis in later development.

ventral lip   figs. 72, 74, 87-90
also see dorsal lip; lateral lip
The margin of the blastopore toward the animal pole and at the ventral side of the embryo; derives from the ventral marginal zone and forms ventral mesoderm in amphibians.

ventral liver bud   *see* posterior liver diverticulum

ventral mesentery   figs. 216, 265
also see dorsal mesocardium; lesser omentum; ventral mesoesophagus
A double layer of splanchnic mesoderm attaching parts of the foregut to the ventral body wall; forms the transient mesocardia, roots of the lungs, lesser omentum, and falciform ligament of the liver.

ventral mesocardium   fig. 169
also see dorsal mesocardium
A temporary mesentery attaching the ventral wall of the heart to the body wall.

ventral mesoderm   figs. 90, 91, 104
also see lateral plate mesoderm
The extension of the lateral plate into the ventral body region; split by the coelom into somatic and splanchnic mesoderm.

ventral mesoesophagus   figs. 256, 257
also see dorsal mesocardium
The ventral mesentery of the esophagus; forms the roots of the lungs and the transient mesocardia.

ventral pancreas   figs. 235, 263-265
also see dorsal pancreas
A ventral evagination of the liver diverticulum (two in birds and amphibians) which grows, branches, and fuses with the dorsal pancreas to form one glandular mass of the adult pancreas.

ventral ramus of spinal nerve   figs.259, 261
also see spinal nerve; the main ventral branch of the spinal nerve trunk; innervates the viscera, body wall, and limbs.

ventral root of spinal nerve   figs. 259, 261, 270, 271
also see spinal nerve; the ventral division of a

V

spinal nerve connecting the trunk of the
nerve to the basal plate of the spinal
cord; composed of motor nerve fibers
arising from neuroblasts in the mantle layer
of the basal plate. *Syn.*: motor root.

ventral vein   fig. 236

Transient veins extending along the ventral
margin of the mesonephros in pig embryos
and draining into the posterior cardinal
veins; subsequently replaced by the
subcardinal veins.

ventricle   figs. 107, 119, 120, 125, 135,
170, 172, 173, 177, 183–187, 197–199,
203, 204, 206–210, 217, 224, 226, 232,
234, 237–239, 259
also see heart; left ventricle; right ventricle
The thick-walled heart chamber that, in the
embryo, receives blood from the atrium
and delivers it under high pressure to
the bulbus cordis; in amniotes, is
partitioned into right and left ventricles
delivering blood to the pulmonary aorta
and ascending aorta, respectively.

ventricular septum   figs. 253, 255, 256
A muscular partition arising from the
posterior wall of the primitive ventricle;
grows anteriorly, fusing with the
endocardial cushion and bulbar septum;
divides the ventricle into right and left
ventricles.

vertebral arteries   figs. 225, 236, 246
A pair of longitudinal vessels extending
anteriorly from the subclavian arteries to
the basilar artery under the
myelencephalon; with the internal
carotids, provides the arterial supply to
the brain; arises from anastomosis of
anterior intersegmental arteries.

vesicular follicle   *see* Graafian follicle

villus cytotrophoblast   figs. 274, 275
also see placental villi
The inner epithelial layer covering
placental villi; cells are separate and may
show mitotic figures; probably form the
overlying syntrophoblast.
Syn.: Langhans cells.

villus syntrophoblast   figs. 273–275
also see peripheral syntrophoblast;
placental barrier; villus cytotrophoblast;

The outer epithelial covering of placental
villi, the part of the chorionic surface and

placental barrier in contact with maternal
blood; a true syncytium; possesses a
brush border on its free surface; partly
underlaid by and derived from the
cytotrophoblast; probable site of synthesis
of placental hormones.

visceral cleft   *see* branchial cleft

visceral groove   *see* branchial groove

visceral pouch   *see* pharyngeal pouch

vitelline arteries   figs. 183, 185, 186,
206–208, 210, 220, 224, 226, 229, 246
The arterial supply of the yolk sac; arise
as ventral branches of the dorsal aortae;
form the superior mesenteric artery and,
in mammals, the coeliac and inferior
mesenteric arteries as well.
Syn.: omphalomesenteric arteries.

vitelline membrane   figs. 16, 27, 44,
146, 188
A membrane enveloping the egg or oocyte;
lies immediately outside the
plasmalemma; formed while the
oocyte is in the ovary and, in some
species, after fertilization, separates from
the egg to form the fertilization membrane.

vitelline plexus   figs. 170, 179–181, 186
A network of small vessels in the yolk sac;
some enlarge to form the vitelline veins
and arteries.

vitelline veins   figs. 107, 118–120, 168,
170–173, 178, 185–187, 198, 199, 206,
217–219, 225–226, 228, 236, 260, 266
also see anterior vitelline veins; posterior
vitelline vein
Vessels that provide initially the venous
return from the yolk sac; proximal ends
fuse, forming successively the atrium, and
sinus venosus of the heart, and the ductus
venosus; also form the hepatic vein,
hepatic sinusoids, and the hepatic portal
vein; distal branches degenerate with the
yolk sac; in amphibians, form around the
yolk endoderm of the midgut.
Syn.: omphalomesenteric veins.

vitelline vessels   figs. 174, 189, 192,
193, 211, 221
also see vitelline arteries; vitelline plexus;
vitelline veins
The blood vessels of the yolk sac; arise
from the blood islands.

wheel organ   fig. 56

In Amphioxus, a dark-staining ring of ciliated epithelium posterior to the oral hood; transports mucous with food toward the mouth and into the pharynx.

wing bud   figs. 203, 207–210, 218, 219, 224–226, 228

The rudiment of the wing; arises as a thickening of somatic mesoderm of the body wall; later bears an ectodermal thickening, the apical ridge; homologous to the foreleg bud and arm bud.

Wolffian body   *see* mesonephros

Wolffian duct   *see* mesonephric duct

X-chromosome   figs. 11, 12, 13

also see leptotene stage

The sex chromosome which usually occurs double in females and single in males, where it may be associated with a Y-chromosome; often exists in a contracted or heteropyknotic state.

yolk   figs. 27, 43, 58, 144, 149, 150

A reserve food mixture within the ovum; in birds formed into yolk spheres up to 100 microns in diameter; composed mainly of lipids and proteins.

yolk endoderm   figs. 95, 101–104, 106, 107, 113, 115, 116, 118, 119

A mass of large yolky cells in the floor of the midgut in amphibians; derives from the vegetal hemisphere; subsequently cells disintegrate and the yolk is absorbed.

yolk plug   figs. 87–91

A mass of large yolky cells filling the blastopore of the amphibian gastrula; derives from the vegetal hemisphere of the the blastula; invaginates to form the yolk endoderm of the neurula.

yolk sac   figs. 174–178, 187–192, 196, 197, 199, 202, 208–211, 215–223, 226–228, 230, 231, 232, 234, 237

also see splanchnopleure

A bag-like extraembryonic membrane formed as an extension of the midgut; in vertebrates with large eggs, encloses and absorbs the yolk; in mammals, is filled with fluid; arises from splanchnopleure and contains vitelline blood vessels; the earliest blood-forming organ; the source of the primordial germ cells; forms a placenta in some elasmobranchs and mammals (pig); degenerates eventually.

yolk stalk   figs. 188, 235

The narrow connection of the yolk sac to the midgut; contains the vitelline arteries and veins.

zona pellucida   figs. 7, 8

A thick membrane containing mucopolysaccharide surrounding the eggs of mammals; called a zona radiata in the ovary when perforated by cytoplasmic processes of the oocyte and follicle cells; penetrated by sperm during fertilization and encloses the embryo during cleavage.

zygotene stage   figs. 2, 143

also see leptotene stage; pachytene stage

The stage of prophase of the first meiotic division when homologous chromosomes pair or synapse; preceded by the leptotene stage and followed by the pachytene.

Photographic Data

EMBRYOLOGICAL PREPARATIONS

All prepared slides photographed for this atlas were selected from the embryological collection of the Department of Biological Sciences, Wayne State University. The slides were commercially produced by several biological supply houses. Only the opaque mounts of chick embryos (figs. 160 and 205) were specially prepared for the atlas. For these two figures, chick embryos were fixed in a saturated solution of mercuric chloride and transferred to 70% alcohol. The embryos were photographed in alcohol. The pig embryo opaque mount (fig. 233) was fixed in Bouin's fluid and preserved in 70% alcohol for photograph. All amphibian material was prepared from *Ran pipiens* embryos; the *Ascaris* was *A. megalocephala*.

EQUIPMENT

Microscope—Bausch and Lomb Dynoptic microscope, Model CBR-9, with achromatic condenser.

Camera—Bausch and Lomb Photomicrographic Camera, Model L.

Illuminators—Bausch and Lomb Research Illuminator with a ribbon filament lamp was used for all photomicrographs. The substage Fluorescent Illuminator was used with the Micro Tessar lens.

Filters—Set of Wratten color filters used to increase or decrease the contrast of the photomicrographs as required.

Enlarger—Omega D2 with a Wollensak lens, 135 mm, $f/4.5$.

PHOTOGRAPHIC SUPPLIES

Films—Kodak 4" x 5" cut film, LS Pan, Pantomic X, Plus X, Ektachrome, Technical Pan 2415

Developers—Kodak D19 and HC110 were used for most plates, but D76 was used to reduce contrast of some whole mount photomicrographs.

Photographic Paper—Kodabromide in contrast surfaces from F1 to F5 as needed.